AF389153

Newborn Screening for Pompe Disease

Newborn Screening for Pompe Disease

Editors

Wuh-Liang Hwu
Yin-Hsiu Chien
Raymond Wang

MDPI • Basel • Beijing • Wuhan • Barcelona • Belgrade • Manchester • Tokyo • Cluj • Tianjin

Editors

Wuh-Liang Hwu
National Taiwan University
Hospital
Taiwan

Yin-Hsiu Chien
National Taiwan University
Hospital
Taiwan

Raymond Wang
Children's Hospital of Orange
County
USA

Editorial Office
MDPI
St. Alban-Anlage 66
4052 Basel, Switzerland

This is a reprint of articles from the Special Issue published online in the open access journal *International Journal of Neonatal Screening* (ISSN 2409-515X) (available at: https://www.mdpi.com/journal/IJNS/special_issues/pompe).

For citation purposes, cite each article independently as indicated on the article page online and as indicated below:

LastName, A.A.; LastName, B.B.; LastName, C.C. Article Title. *Journal Name* **Year**, *Volume Number*, Page Range.

ISBN 978-3-0365-0580-0 (Hbk)
ISBN 978-3-0365-0581-7 (PDF)

Cover image courtesy of the American Academy of Pediatrics.

Contents

About the Editors

Wuh-Liang Hwu is a Professor in the Department of Pediatrics at the National Taiwan University Hospital (NTUH) in Taiwan. Professor Hwu completed his medical and PhD degrees at the College of Medicine, National Taiwan University. He completed his residency at NTUH. He has also done a fellowship at the Department of Genetics at Johns Hopkins University and was a Visiting Scientist at the Department of Medical Genetics at Mayo Clinic. He was the Department Head of Medical Genetics at NTUH from 2006 to 2012 and the Inaugural President of the Taiwan Human Genetics Society from 1999 to 2002. He was a Board Member of the Taiwan Foundation for Rare Disorders. Professor Hwu set up the Newborn Screening Program for Pompe Disease in Taiwan, one of the first in the world, and has dedicated much of his research and clinical effort to the diagnosis, management, and gene therapy of patients with rare genetic diseases. His recent interests include employing next-generation sequencing (NGS) and artificial intelligence to facilitate the diagnosis of genetic diseases and related conditions.

Yin-Hsiu Chien is a Clinical Professor at the Department of Pediatrics at the National Taiwan University, Taipei, Taiwan, and an Attending Physician of the Department of Medical Genetics and Pediatrics at the National Taiwan University Hospital. She is the Director of the newborn screening center at the National Taiwan University Hospital, which routinely screens around one-third of newborns in Taiwan. Her team, led by Dr Wuh-Liang Hwu, is devoted to the diagnosis and treatment of lysosomal storage diseases, neurotransmitter deficiency, and neuromuscular disorders. Dr Chien's current work focuses on the early diagnosis and improvement of treatment for Pompe disease, AADC deficiency, spinal muscular atrophy, and other lysosomal storage disorders.

Raymond Wang MD, is the Campbell Foundation of Caring Director of the Multidisciplinary Lysosomal Storage Disorder Program at the Children's Hospital of Orange County in Orange, California. He is a member of the faculty at the Division of Metabolic Disorders at CHOC Children's Specialists and an Associate Clinical Professor of Pediatrics at UC Irvine Medical School. He oversees clinical care for patients with all lysosomal storage disorders, as well as clinical trials and translational research with the goal of providing patients access to life-saving therapies, and is involved with developing next-generation therapies in the laboratory using gene transfer and CRISPR genome editing.

International Journal of
Neonatal Screening

Editorial

Development of Newborn Screening for Pompe Disease

Wuh-Liang Hwu [1,2,*] and Yin-Hsiu Chien [1,2]

[1] Department of Pediatrics, National Taiwan University Hospital, Taipei 10041, Taiwan; chienyh@ntu.edu.tw
[2] Department of Medical Genetics, National Taiwan University Hospital, Taipei 10041, Taiwan
* Correspondence: hwuwlntu@ntu.edu.tw

Received: 28 December 2019; Accepted: 22 January 2020; Published: 24 January 2020

Pompe disease is an inborn error of lysosomal degradation of glycogen. The responsible enzyme is acid alpha-glucosidase (GAA). In the severe form of the disease, or infantile-onset Pompe disease (IOPD), weakness in the skeletal and cardiac muscles soon leads to both respiratory and cardiac failure, and death usually occurs before the age of one year. In 2006, the US Food and Drug Administration (FDA) approved Myozyme, the first drug for Pompe disease [1]. My group at the National Taiwan University Hospital (NTUH) joined the phase III clinical trial. A few patients in Taiwan were closely followed before and after they were enrolled in the trial, or an expanded-use program [2]. One patient underwent two muscle biopsies before starting treatment, and had another two biopsies thereafter. We observed progressive degeneration of the patient's muscles. Debris-like materials filled almost all of the spaces in the myocytes in the last biopsy. Clinically, the patient gradually lost all motor abilities. We were shocked by the irreversible nature of the skeletal muscle pathologies and had hoped that the patient could have been treated earlier.

However, it is difficult for caregivers to detect muscle weakness in young infants. For example, in the Chinese culture, babies are tightly swaddled in blankets, so there are few opportunities to observe the motor developments of young infants until they can be held on their caregivers' shoulders at an age of around 3–4 months. More often, patients with IOPD are diagnosed incidentally during respiratory infection at ages 4–5 months. Unfortunately, when signs of weakness appear, muscle damage due to glycogen storage is already extensive and irreversible. Therefore, we studied newborn screening for Pompe disease [3].

In 2006, the screening laboratory at NTUH already had experience with the in-house development of tandem mass newborn screening, and the biochemical laboratory had conducted enzymatic diagnoses of Pompe disease for more than 10 years. Two techniques, critical for Pompe disease newborn screening, were also developed at that time. The first, invented by Dr. Nestor Chamoles [4], measures lysosomal enzyme activities in punches from dried blood spots (DBS). The second was the discovery of an inhibitor, acarbose, of maltase-glucoamylase, another acid glucosidase abundant in the leukocytes [5]. Using these two methods, we measured GAA activity in DBS, eluted via overnight incubation with a fluorescence substrate. Funding is another critical element of Pompe screening. Luckily, we were able to persuade Dr. Joan Keutzer from Genzyme about the necessity of newborn screening for Pompe disease, and our ability to perform the requisite tests. Thanks to the excellent work of my senior laboratory scientist, Shu-Chuan (Sara) Chiang, and my successor, Professor Yin-Hsiu Chien, we established the protocol and proved it step-by-step. We demonstrated the screening results [6], defined the molecular epidemiology for GAA pseudodeficiency [7], proved the outcome of IOPD patients detected by screening [8], explored the features of less severe patients detected by screening [9], and then revised the screening algorithm [10]. We also added Fabry disease to Pompe disease screening using a fluorescence substrate [11]. However, true multiplex newborn screening was made possible only after the development of tandem mass substrates by Dr. Michael Gelb at the University of Washington [5,12].

Soon after, several pilot programs, including programs in Italy, Australia, Japan, Korea, USA, and Hungary, have tested the feasibility of Pompe newborn screening and understood the incidence of Pompe disease and the impact of the disease. After the Discretionary Advisory Committee on Heritable Disorders in Newborns and Children (DACHDNC) added Pompe disease to the Recommended Uniform Screening Panel (RUSP) in 2013, the spread of Pompe disease newborn screening increased. However, challenges remain, including sensitivity and specificity of the assays, management of pseudodeficiency, time and method to treat IOPD patients detected by screening, immunomodulation, and management of later-onset Pompe disease patients discovered by screening. In this special issue of the *International Journal of Neonatal Screening*, global experiences with Pompe disease newborn screening were pooled to enhance the understanding of screening and improve the outcomes of patients affected by Pompe disease.

Conflicts of Interest: Y.-H.C. has served on advisory boards for Amicus Therapeutics and Sanofi Genzyme, undertaken contracted research for Sanofi Genzyme, received honoraria, consulting fees, and travel expenses from Sanofi Genzyme. W.-L.H. has served on advisory boards for Audentes and Sanofi Genzyme, undertaken contracted research for Sanofi Genzyme, received honoraria, consulting fees, and travel expenses from Sanofi Genzyme.

References

1. Kishnani, P.S.; Corzo, D.; Nicolino, M.; Byrne, B.; Mandel, H.; Hwu, W.L.; Leslie, N.; Levine, J.; Spencer, C.; McDonald, M.; et al. Recombinant human acid α-glucosidase: Major clinical benefits in infantile-onset Pompe disease. *Neurology* **2007**, *68*, 99–109. [CrossRef] [PubMed]
2. Chien, Y.H.; Lee, N.C.; Peng, S.F.; Hwu, W.L. Brain development in infantile-onset Pompe disease treated by enzyme replacement therapy. *Pediatr. Res.* **2006**, *60*, 349–352. [CrossRef] [PubMed]
3. Kemper, A.R.; Hwu, W.L.; Lloyd-Puryear, M.; Kishnani, P.S. Newborn screening for Pompe disease: Synthesis of the evidence and development of screening recommendations. *Pediatrics* **2007**, *120*, e1327–e1334. [CrossRef] [PubMed]
4. Chamoles, N.A.; Blanco, M.; Gaggioli, D.; Casentini, C. Tay-Sachs and Sandhoff diseases: Enzymatic diagnosis in dried blood spots on filter paper: Retrospective diagnoses in newborn-screening cards. *Clin. Chim. Acta* **2002**, *318*, 133–137. [CrossRef]
5. Li, Y.; Scott, C.R.; Chamoles, N.A.; Ghavami, A.; Pinto, B.M.; Turecek, F.; Gelb, M.H. Direct multiplex assay of lysosomal enzymes in dried blood spots for newborn screening. *Clin. Chem.* **2004**, *50*, 1785–1796. [CrossRef] [PubMed]
6. Chien, Y.H.; Chiang, S.C.; Zhang, X.K.; Keutzer, J.; Lee, N.C.; Huang, A.C.; Chen, C.A.; Wu, M.H.; Huang, P.H.; Tsai, F.J.; et al. Early detection of Pompe disease by newborn screening is feasible: Results from the Taiwan screening program. *Pediatrics* **2008**, *122*, e39–e45. [CrossRef] [PubMed]
7. Labrousse, P.; Chien, Y.H.; Pomponio, R.J.; Keutzer, J.; Lee, N.C.; Akmaev, V.R.; Scholl, T.; Hwu, W.L. Genetic heterozygosity and pseudodeficiency in the Pompe disease newborn screening pilot program. *Mol. Genet. Metab.* **2010**, *99*, 379–383. [CrossRef] [PubMed]
8. Chien, Y.H.; Lee, N.C.; Thurberg, B.L.; Chiang, S.C.; Zhang, X.K.; Keutzer, J.; Huang, A.C.; Wu, M.H.; Huang, P.H.; Tsai, F.J.; et al. Pompe disease in infants: Improving the prognosis by newborn screening and early treatment. *Pediatrics* **2009**, *124*, e1116–e1125. [CrossRef] [PubMed]
9. Chien, Y.H.; Lee, N.C.; Huang, H.J.; Thurberg, B.L.; Tsai, F.J.; Hwu, W.L. Later-onset Pompe disease: Early detection and early treatment initiation enabled by newborn screening. *J. Pediatr.* **2011**, *158*, 1023–1027. [CrossRef] [PubMed]
10. Chiang, S.C.; Hwu, W.L.; Lee, N.C.; Hsu, L.W.; Chien, Y.H. Algorithm for Pompe disease newborn screening: Results from the Taiwan screening program. *Mol. Genet. Metab.* **2012**, *106*, 281–286. [CrossRef] [PubMed]

11. Hwu, W.L.; Chien, Y.H.; Lee, N.C.; Chiang, S.C.; Dobrovolny, R.; Huang, A.C.; Yeh, H.Y.; Chao, M.C.; Lin, S.J.; Kitagawa, T.; et al. Newborn screening for Fabry disease in Taiwan reveals a high incidence of the later-onset GLA mutation c.936+919G>A (IVS4+919G>A). *Hum. Mutat.* **2009**, *30*, 1397–1405. [CrossRef] [PubMed]
12. Spacil, Z.; Tatipaka, H.; Barcenas, M.; Scott, C.R.; Turecek, F.; Gelb, M.H. High-throughput assay of 9 lysosomal enzymes for newborn screening. *Clin. Chem.* **2013**, *59*, 502–511. [CrossRef] [PubMed]

International Journal of
Neonatal Screening

Review

Establishing Pompe Disease Newborn Screening: The Role of Industry

Joan M. Keutzer [†]

Sanofi Genzyme, Cambridge, MA 02142, USA; keutzer@verizon.net
† Retired.

Received: 22 May 2020; Accepted: 3 July 2020; Published: 5 July 2020

Abstract: When clinical trials for enzyme replacement therapy for Pompe disease commenced, a need for newborn screening (NBS) for Pompe disease was recognized. Two methods for NBS for Pompe disease by measuring acid α-glucosidase in dried blood spots on filter paper were developed in an international collaborative research effort led by Genzyme. Both methods were used successfully in NBS pilot programs to demonstrate the feasibility of NBS for Pompe disease. Since 2009, all babies born in Taiwan have been screened for Pompe disease. Pompe disease was added to the Recommended Uniform (Newborn) Screening Panel in the United States in 2015. NBS for Pompe disease is possible because of the unprecedented and selfless collaborations of countless international experts who shared their thoughts and data freely with the common goal of establishing NBS for Pompe disease expeditiously.

Keywords: newborn screening; Pompe disease; acid α-glucosidase

1. Introduction

We frequently hear that "it takes a village" to get things done. In the case of Pompe disease newborn screening (NBS), "it took the world". NBS for Pompe disease resulted from an unprecedented international collaboration among patients with Pompe disease, the Pompe disease medical and scientific communities, government agencies, and industry. Starting in 1999, the baton was passed between research groups who acted urgently to generate the data required, develop methods for newborn screening for Pompe disease, and begin to establish newborn screening for Pompe disease.

In 1998, Genzyme (now Sanofi Genzyme) initiated a clinical development program for enzyme replacement therapy (ERT) for Pompe disease. During meetings to plan the clinical trials, physicians voiced concerns about the ability of patients with Pompe disease to get an early and accurate diagnosis. They cautioned that if the deficiencies in diagnostic testing were not addressed, that Genzyme would not be able to conduct successful clinical trials with ERT in Pompe disease because some symptoms would be irreversible. Thus, diagnostic delays could result in the inability to determine if ERT for Pompe disease was a viable treatment option. Patients with later-onset Pompe disease (LOPD) referred to the test as unreliable and invasive. Some patients told Genzyme that they had three muscle biopsies before they received a diagnosis. Parents of patients with infantile-onset Pompe disease (IOPD) mentioned waiting two grueling months after the collection of a skin biopsy to get a diagnosis. There was no assay that could use blood samples to diagnose Pompe disease.

In response, I assembled a team of Genzyme colleagues to determine if there was a role that we could play in improving diagnostics for Pompe disease. The team consisted of members from Research and Development, Genzyme Diagnostics, Genzyme Genetics, Regulatory Affairs, and the Pompe disease Clinical Development Team. The team interviewed multiple Pompe disease stakeholders, including patients, physicians, and diagnostic laboratories. Ultimately, Henri Termeer, the CEO of Genzyme, decided to sponsor international research to develop assays useful for NBS, diagnosis using blood samples, patient monitoring, and genotyping for Pompe disease. This decision was not supported unanimously internally. Some felt very strongly that it created a conflict of interest if a company was developing therapeutic and diagnostic testing for the same disease. Several healthy conversations about our role in patient diagnosis ensued and led to a commitment to keep our involvement in assay development scientific and make the resultant intellectual property and methodologies available to all. Much of the work was co-sponsored or sponsored by Genzyme. In 2003, Genzyme established an R&D lab to optimize, validate, and transfer methodology. The team was active through 2016.

2. Early Attempts to Develop Newborn Screening Assays

We hoped that NBS for Pompe disease could replicate what was done for other enzyme deficiencies, like PKU, and measure the accumulation of the enzyme's substrate in dried bloodspots on filter paper (DBS) in a multiplex assay that could screen for multiple treatable lysosomal storage disorders (LSDs) simultaneously. In 2000, we started to collaborate with Sarah Young and David Millington on work they had initiated earlier. They developed the urine Hex4 assay, which is an indirect measure of elevated glycogen. The assay did not translate into an assay for DBS, but it is useful for monitoring patients with Pompe disease [1].

We knew that the measurement of acid α-glucosidase enzyme activity in mixed leukocytes and whole blood was complicated by the presence of another alpha-glucosidase, maltase glucoamylase (MGA), which is active at an acid pH and masks the deficiency of acid α-glucosidase; therefore, we sponsored the search for an inhibitor of MGA so that acid α-glucosidase could be quantitated in blood. For the first two years, no one found a useful inhibitor, but labs in Europe, Asia, and the United States kept trying. At International Conference on Inborn Errors of Metabolism (ICIEM) in Cambridge, UK in September 2000, Nestor Chamoles from the Laboratorio Chamoles in Buenos Aires, Argentina presented a poster that demonstrated that it was possible to measure the lysosomal enzyme alpha-L-iduronidase in DBS and differentiate DBS from patients with MPS I from DBS from obligate heterozygotes and healthy individuals [2]. We encouraged Chamoles to try to create an assay for acid α-glucosidase in DBS using 4-methylumbelliferyl-α-D-glucopyranoside (4-MUG) as the substrate.

3. Measuring Acid α-Glucosidase in Dried Bloodspots on Filter Paper

We knew that NBS was relying more heavily on multiplex assays, so we continued to search for technologies that could be used in a multiplex assay for LSDs. Our assumption was that like the other multiplex assays in NBS, the assay would measure analytes (substrates of the missing enzymes) in DBS, but that changed to wanting to measure several enzyme activities in a multiplex assay. At the same ICIEM meeting in Cambridge, UK in September 2000, Frantisek Turecek, C. Ronald Scott, and Michael H. Gelb from the University of Washington presented an assay that measured acid sphingomyelinase and β-galactocerebrosidase in skin fibroblast homogenates for clinical laboratory diagnosis [3]. The assay used novel biotinylated substrate conjugates, purification of the products on streptavidin agarose beads, and quantification of products using electrospray ionization mass spectrometry with stable-isotope-labeled internal standards that were chemically identical to the products of the enzymatic reactions. We invited them to join the efforts and their initial goal was to develop a multiplex assay for six lysosomal storage disorders Pompe disease, MPS I, Fabry disease, Gaucher disease, Krabbe disease, and Niemann–Pick disease types A and B.

In 2003, Gabriella Niizawa from the Chamoles lab developed a fluorometric assay using 4-methylumbelliferyl-α-D-glucopyranoside (4-MUG) as the substrate that could differentiate DBS from patients with IOPD from obligate heterozygotes and healthy individuals. The method used maltose to inhibit MGA since MGA has a higher affinity for maltose than acid α-glucosidase [4]. Chamoles invited X. Kate Zhang and me to his lab to run the assay and in June 2003, and allowed us to take his protocol into the Genzyme LSD assay development lab so we could try to optimize the protocol and adapt it from one using 1.5 mL microcentrifuge tubes to one that uses in 96-well plates and could be automated. Our activities were limited by the paucity of samples from patients previously diagnosed with Pompe disease.

In September 2003, Genzyme received a phone call from a father from Peru. He wanted to know how to access treatment for Pompe disease because his 4-month-old baby was suspected of having the disease because her brother died from Pompe disease. The baby did not have a confirmed diagnosis. We talked to Chamoles who said he was not ready to use the DBS assay for suspected cases but would test and report a result if Genzyme tested duplicate samples blinded and the results concurred. We requested duplicate DBS from the baby, the mother, the father, and someone unrelated. Peru did not have an NBS program, so we had to use a courier to deliver DBS cards to the clinic. Chamoles and Genzyme had similar results. Samples from both parents and the unrelated donor were in the normal range and the sample from the baby was consistent with a diagnosis of Pompe disease. The child and her family relocated to Durham, NC to participate in the ongoing clinical trial at Duke University Medical Center. The diagnosis of Pompe disease was confirmed at Duke using an assay in fibroblasts. This patient was the first identified with Pompe disease using the new blood-based assay and we gained confidence in the methodology.

Chamoles, Gelb, and the Genzyme R&D Team met at ICIEM in Brisbane in September 2003 where Chamoles presented data on maltose inhibition of MGA. We agreed to meet as soon as possible to discuss sharing lab methods. At the time, Gelb did not have a DBS assay for Pompe disease. In January 2004, Genzyme hosted a mini symposium to generate a collaborative research agenda. At that meeting, Gelb and Turecek revealed that they identified 80 μM acarbose (a drug used to control blood glucose levels in type 2 diabetes) to inhibit MGA [5]. However, all three labs had issues with sample availability. Chamoles had a limited number of samples because he collected DBS from patients undergoing diagnostic evaluation for Pompe disease and from their parents who were considered obligate carriers, but he shared them. Genzyme agreed that they would adapt both methods that used microcentrifuge tubes to 96-well dishes so that the relative effectiveness of maltose and acarbose could be studied more easily.

4. Optimizing the Methods

During the adaptation, optimization and validation activities led by Helmut Kallwass, the Genzyme R&D team, determined that acarbose was superior to maltose in the assay and that 8 μM acarbose was superior to 80 μM acarbose. Although 8 μM and 80 μM acarbose selectively blocked MGA over acid α-glucosidase, use of 8 uM acarbose inhibited less acid α-glucosidase (i.e., acarbose does have finite affinity for acid α-glucosidase but has higher affinity for MGA). However, we thought it was important to compare the DBS assay to the fibroblast assay in matched samples. Fortunately, in anticipation of evaluating the DBS assay, the R&D group at Duke University under the leadership of Deeksha Bali collected DBS from some IOPD patients for which they had fibroblast cultures. Bali's team saw discrimination between patients with IOPD, obligate heterozygotes, and controls using 8 μM acarbose and the results using acarbose compared well with those using the skin fibroblast assay in the patients [6].

The Genzyme Clinical Development Team assisted in sample collection. In 2003, the informed consent and protocol for the pivotal Phase 3 clinical trial for patients with IOPD (A Study of the Safety and Efficacy of rhGAA in Patients with Infantile-onset Pompe Disease (NCT00059280)) were amended to include collecting DBS for developing DBS enzyme activity assays for acid α-glucosidase. In 2004, a Prospective, Observational Study in Patients with Late-Onset Pompe Disease (NCT00077662) included an optional request for DBS samples. In 2005, A Placebo-Controlled Study of Safety and Effectiveness of Myozyme (alglucosidase alfa) in Patients with Late-Onset Pompe Disease called the Late-Onset Treatment Study [LOTS] (NCT00158600) included an optional request for DBS samples during screening. There was virtually 100% patient participation in donating DBS in the three clinical trials. Without these samples, we could not have progressed in developing NBS or blood-based diagnostic testing for Pompe disease.

The DBS assay consistently differentiated the 24 previously diagnosed IOPD and 100 patients screened for LOTS from controls. Testing DBS samples from 61 patients from LOPOS previously diagnosed with LOPD revealed that 58 had results consistent with a diagnosis of Pompe disease and three had results in the normal range. Measuring acid α-glucosidase enzyme activity in fibroblasts and acid α-glucosidase gene (*GAA*) sequence analysis confirmed that the three patients with a neuromuscular disease were misdiagnosed with Pompe disease [7]. Four fibroblast samples had ≤1% residual enzyme activity so results from fibroblasts cannot be used to predict disease phenotype in cases identified by newborn screening. The results in the DBS assay were similar for IOPD and LOPD, so the DBS assay cannot be used to predict disease phenotype.

It is worth noting that all participating DBS collection sites were trained on how to make DBS via conference call and none of the DBS received from the clinical sites were unsatisfactory for use in assay development; this suggested that it might be feasible to use DBS samples for clinical diagnosis. DBS sampling permits access to testing for metabolic diseases in remote areas; DBS samples can be conveniently collected and shipped through the mail to a distant laboratory for analysis.

5. Early Experience with the Acid α-Glucosidase Enzyme Assay

Paul (Wuh-Liang) Hwu and Nancy (Yin-Hsiu) Chien invited Chamoles and I to a meeting at the Asian Regional International NBS Meeting in September 2004. They proposed that they were uniquely situated to conduct the first large-scale Pompe disease NBS pilot. They described a plan to treat those diagnosed with IOPD immediately and follow the patients' outcomes using the protocols identical to the pivotal alglucosidase alfa clinical trial in IOPD, since they were a clinical trial site. We transferred that methodology from our Genzyme R&D lab to the National Taiwan University Hospital (NTUH) NBS Center in October 2004. NTUH introduced the word's first Pompe disease NBS pilot program at the end of 2005 [8]. Yuan-Tsong Chen from Academia Sinica in Taipei, Taiwan collaborated in the design and execution of the program.

Initially, the NTUH lab ran three separate fluorometric assays to measure acid α-glucosidase, neutral α-glucosidase (NAG), and maltase-glucoamylase (MGA) activities. NAG and MGA were used to differentiate cases of Pompe disease from false positive cases. As the pilot progressed and the lab had more experience, cautious and methodical adjustments of assay cut-offs were used to minimize the risk of false positive and false negative test results. During the pilot, NTUH identified the high prevalence of the *GAA* pseudodeficiency allele (p.G576S) and the increase in phenotype severity when the polymorphism is in cis [9]. Since 2009, all newborns in Taiwan have been screened for Pompe disease. In 2015, NTUH adopted the tandem mass spectrometry-based multiplex LSD enzyme assay.

Today, it seems obvious that the DBS assay with acarbose is useful in diagnosing Pompe disease since all assays for Pompe disease using samples from blood include acarbose, but previous experiences with Pompe disease diagnoses and misdiagnoses left many skeptical of the new method in its early days. In December 2006, Genzyme sponsored a meeting of The Pompe Disease Diagnostic Working Group in London to establish a consensus regarding the application of these new assays for the laboratory diagnosis of Pompe disease. The Working Group consisted of scientists, geneticists, and clinicians working in the field of Pompe disease. The group agreed that cultured skin fibroblasts had been the gold standard, but was rapidly being replaced by assays in blood samples because they are less invasive, more rapid, and are easier to standardize [10]. The meeting was an important step in bringing laboratories and clinicians together to agree that the DBS assay using acarbose to inhibit MGA was useful to get a presumptive positive diagnoses for Pompe disease and that at least one secondary positive test (for example, *GAA* sequence analysis) would support a biochemical diagnosis of Pompe disease. Genzyme's R&D team made the DBS assay protocol available to anyone who requested it and trained labs at Genzyme and remotely as requested. This hastened the adoption of the fluorescent assay to measure acid α-glucosidase enzyme activity and increased the number of patients being diagnosed with Pompe disease worldwide.

While the team at NTUH was using the Genzyme R&D adaptation of the Chamoles lab assay in their NBS program, the Genzyme R&D team that was led by Zhang was adapting the tandem mass spectrometry-based assay for multiple LSD enzymes from the University of Washington [5] to a high-throughput method. However, Genzyme did not have any experience in NBS, so we asked each NBS lab in the United States to answer a survey about the published assay and its suitability for use in NBS. For each survey returned, Genzyme made a donation to the Association of Public Health Laboratories. Based on the responses from NBS labs, Genzyme transformed the assay into a robust high throughput 96-well plate format assay using a robotic liquid handling system for sample transfer and minimizing the detergent burden on the mass spectrometer by replacing Triton X-100 with CHAPS. We addressed the environmental concerns of the NBS labs by replacing chloroform with ethyl acetate [11]. We created a reliable supply of the substrates and internal standards manufactured in our GMP facility using validated manufacturing processes.

We facilitated the use of the assay in NBS through a ten year (2006–2016) donation and grant agreement with the Centers for Disease Control and Prevention (CDC) Foundation so that (1) the reagents could be distributed without charge to NBS labs before they were commercially available, (2) NBS labs could be trained on lysosomal storage disorder NBS methods, and (3) quality control and external quality assurance samples were available for use in NBS [12]. The reagents were manufactured at Genzyme Pharmaceuticals in Liestal, Switzerland and tested for quality and distributed to the CDC by Genzyme Diagnostics in Framingham, Massachusetts.

6. Pompe Disease and the Recommended Uniform (Newborn) Screening Panel

In May 2006, Priya Kishnani nominated Pompe disease for inclusion in the Recommended Uniform (Newborn) Screening Panel (RUSP) in the United States. In October 2008, the nomination was denied. The Advisory Committee recommended additional studies before the condition could be re-nominated. The response to the nomination stated that "the initial test specificity for Pompe disease alone should be improved in comparison to the data shown by the Taiwan study, further evaluation of alternative screenings methods that could be multiplexed to target additional conditions (for example, other Lysosomal Storage Disease conditions) in order to decrease the burden on public health laboratories, and a standardized method of diagnosis after a positive newborn screen is required [13].

The Pompe disease community and Genzyme were disappointed by the response, but Genzyme remained staunch and committed to two additional efforts: funding an NBS study in Washington state and sponsoring an expert group to generate guidance for NBS for Pompe disease for practitioners around the world.

Fortuitously, the team in Washington had just developed a true triplex enzyme assay for Pompe disease, Fabry disease, and MPS I in which the enzymatic activities of acid α-glucosidase, α-galactosidase A, and α-L-iduronidase were quantified in DBS using a single assay buffer [14]. The validated triplex assay was transferred expeditiously to the Washington State NBS Lab and used to measure enzyme activities in approximately 110,000 DBS that had been stored at 18 °C for 8–10 months after collection. The word approximately is used because the study started before the reagents for Fabry disease and MPS I were ready for distribution by the CDC and results were not reported for tests done using research grade reagents, so the number of results reported vary by disease [15]. The study was anonymous but confirmed the diagnosis by DNA sequencing using a duplicate punch from the same sample. The positive predictive values for Pompe disease, Fabry disease, and MPS I were 0.24, 0.43, and 0.33, respectively. The false positive rates were 1/8600, 1/12,100, and 1/17,500, respectively.

The Washington State NBS data generated a lot of hope in the MPS I and Pompe disease communities. In late 2011, cross-functional teams of patient advocacy groups, diagnostic experts, disease experts, and NBS experts were established for MPS I and Pompe disease. MPS I and Pompe disease were nominated for inclusion in the RUSP in 2012; and the nominations were approved for formal data review. Evidence review for Pompe disease preceded evidence review for MPS I because the Advisory Committee budget could not support parallel evidence reviews. In 2013, the Advisory Committee recommend adding Pompe Disease to the RUSP. In 2015, Pompe disease was added to the RUSP [16]. In 2015, the Advisory Committee recommended adding MPS I to the RUSP. In 2016, MPS I was added to the RUSP [16].

Recognizing the importance of consistency in NBS, Sanofi Genzyme facilitated and provided financial support for the meeting of the Pompe Disease NBS Working Group, which was led by Hwu and Kishnani to discuss and develop a general guidance document for NBS for Pompe disease for practitioners around the world. The guidance was published in a supplement in Pediatrics [17]. Sub-teams from the Working Group authored guidelines for NBS for Pompe Disease; the initial evaluation of patients after positive NBS and recommended algorithms leading to a confirmed diagnosis of Pompe disease; management of confirmed newborn-screened patients with Pompe disease across the disease spectrum; and the role of genetic counseling in Pompe disease after patients are identified through NBS.

NBS for Pompe disease is possible because of the unprecedented and selfless collaborations of countless international experts who shared their thoughts and data without delay so that the next step in the process could be initiated during the lag required for preparation of publications. NBS for Pompe disease was facilitated by the financial, organizational, and technical contributions from Genzyme and Sanofi Genzyme.

Funding: This research received no external funding.

Acknowledgments: The author is grateful to each patient with Pompe disease who donated samples for these studies and to the numerous physicians, scientists, and Genzyme colleagues who participated in all aspects of this work. Their trust was empowering. Henri Termeer is recognized for his unwavering support and pioneering vision, and Nestor Chamoles for his passion and deep commitment. Please note that I referenced the first paper for each major milestone, but recognize that these efforts have culminated in many, many more publications.

Conflicts of Interest: The author was previously a full-time employee of Sanofi Genzyme (formerly Genzyme).

References

1. An, Y.; Young, S.P.; Kishnani, P.S.; Millington, D.S.; Amalfitano, A.; Corz, D.; Chen, Y.T. Glucose tetrasaccharide as a biomarker for monitoring the therapeutic response to enzyme replacement therapy for Pompe disease. *Mol. Genet. Metab.* **2005**, *85*, 247–254. [CrossRef] [PubMed]
2. Chamoles, N.A.; Blanco, M.; Gaggioli, D. Diagnosis of alpha-L-iduronidase deficiency in dried blood spots on filter paper: The possibility of newborn diagnosis. *Clin. Chem.* **2001**, *47*, 780–781. [CrossRef] [PubMed]
3. Zhou, X.; Turecek, F.; Scott, C.R.; Gelb, M.H. Quantification of cellular acid sphingomyelinase and galactocerebroside beta-galactosidase activities by electrospray ionization mass spectrometry. *Clin. Chem.* **2001**, *47*, 874–881. [CrossRef] [PubMed]
4. Chamoles, N.A.; Niizawa, G.; Blanco, M.; Gaggioli, D.; Casentini, C. Glycogen storage disease type II: Enzymatic screening in dried blood spots on filter paper. *Clin. Chim. Acta.* **2004**, *347*, 97–102. [CrossRef] [PubMed]
5. Li, Y.; Scott, C.R.; Chamoles, N.A.; Ghavami, A.; Pinto, B.M.; Turecek, F.; Gelb, M.H. Direct multiplex assay of lysosomal enzymes in dried blood spots for newborn screening. *Clin. Chem.* **2004**, *50*, 1785–1796. [CrossRef] [PubMed]
6. Zhang, H.; Kallwass, H.; Young, S.P.; Carr, C.; Dai, J.; Kishnani, P.S.; Millington, D.S.; Keutzer, J.; Chen, Y.T.; Bali, D. Comparison of maltose and acarbose as inhibitors of maltase-glucoamylase activity in assaying acid alpha-glucosidase activity in dried blood spots for the diagnosis of infantile Pompe disease. *Genet. Med.* **2006**, *8*, 302–306. [CrossRef] [PubMed]
7. Kallwass, H.; Carr, C.; Gerrein, J.; Titlow, M.; Pomponio, R.; Bali, D.; Dai, J.; Kishnani, P.; Skrinar, A.; Corzo, D.; et al. Rapid diagnosis of late-onset Pompe disease by fluorometric assay of alpha-glucosidase activities in dried blood spots. *Mol. Genet. Metab.* **2007**, *90*, 449–452. [CrossRef] [PubMed]
8. Chien, Y.H.; Chiang, S.C.; Zhang, X.K.; Keutzer, J.; Lee, N.C.; Huang, A.C.; Chen, C.A.; Wu, M.H.; Huang, P.H.; Tsai, F.J.; et al. Early detection of Pompe disease by newborn screening is feasible: Results from the Taiwan screening program. *Pediatrics* **2008**, *122*, e39–e45. [CrossRef] [PubMed]
9. Chiang, S.C.; Hwu, W.L.; Lee, N.C.; Hsu, L.W.; Chien, Y.H. Algorithm for Pompe disease newborn screening: Results from the Taiwan screening program. *Mol. Genet. Metab.* **2012**, *106*, 281–286. [CrossRef] [PubMed]

10. Pompe Disease Diagnostic Working Group; Winchester, B.; Bali, D.; Bodamer, O.A.; Caillaud, C.; Christensen, E.; Cooper, A.; Cupler, E.; Deschauer, M.; Fumić, K.; et al. Methods for a prompt and reliable laboratory diagnosis of Pompe disease: Report from an international consensus meeting. *Mol. Genet. Metab.* **2008**, *93*, 275–281. [CrossRef]

11. Zhang, X.K.; Elbin, C.S.; Chuang, W.L.; Cooper, S.K.; Marashio, C.A.; Beauregard, C.; Keutzer, J.M. Multiplex enzyme assay screening of dried blood spots for lysosomal storage disorders by using tandem mass spectrometry. *Clin. Chem.* **2008**, *54*, 1725–1728. [CrossRef] [PubMed]

12. De Jesus, V.R.; Zhang, X.K.; Keutzer, J.; Bodamer, O.A.; Mühl, A.; Orsini, J.J.; Caggana, M.; Vogt, R.F.; Hannon, W.H. Development and evaluation of quality control dried blood spot materials in newborn screening for lysosomal storage disorders. *Clin. Chem.* **2009**, *55*, 158–164. [CrossRef] [PubMed]

13. Advisory Committee on Heritable Disorders in Newborns and Children's response to Dr Kishnani. Available online: https://www.hrsa.gov/sites/default/files/hrsa/advisory-committees/heritable-disorders/rusp/previous-nominations/pompe-letter-from-committee-2008.pdf (accessed on 19 May 2020).

14. Duffey, T.A.; Bellamy, G.; Elliott, S.; Fox, A.C.; Glass, M.; Turecek, F.; Gelb, M.H.; Scott, C.R. A tandem mass spectrometry triplex assay for the detection of Fabry, Pompe, and mucopolysaccharidosis-I (Hurler). *Clin. Chem.* **2010**, *56*, 1854–1861. [CrossRef] [PubMed]

15. Scott, C.R.; Elliott, S.; Buroker, N.; Thomas, L.I.; Keutzer, J.; Glass, M.; Gelb, M.H.; Turecek, F. Identification of infants at risk for developing Fabry, Pompe, or mucopolysaccharidosis-I from newborn blood spots by tandem mass spectrometry. *J. Pediatr.* **2013**, *163*, 498–503. [CrossRef] [PubMed]

16. Advisory Committee on Heritable Disorders in Newborns and Children: Previously Nominated Conditions. Available online: https://www.hrsa.gov/advisory-committees/heritable-disorders/rusp/previous-nominations.html (accessed on 19 May 2020).

17. Newborn Screening, Diagnosis, and Treatment for Pompe Disease. Available online: https://pediatrics.aappublications.org/content/140/Supplement_1 (accessed on 19 May 2020).

International Journal of
Neonatal Screening

Review

Newborn Screening for Pompe Disease

Takaaki Sawada, Jun Kido * and Kimitoshi Nakamura

Department of Pediatrics, Graduate School of Medical Sciences, Kumamoto University, Kumamoto 860-8556, Japan; sawada.takaki@kuh.kumamoto-u.ac.jp (T.S.); nakamura@kumamoto-u.ac.jp (K.N.)
* Correspondence: kidojun@kuh.kumamoto-u.ac.jp; Tel.: +81-96-373-5191

Received: 19 January 2020; Accepted: 2 April 2020; Published: 5 April 2020

Abstract: Glycogen storage disease type II (also known as Pompe disease (PD)) is an autosomal recessive disorder caused by defects in α-glucosidase (AαGlu), resulting in lysosomal glycogen accumulation in skeletal and heart muscles. Accumulation and tissue damage rates depend on residual enzyme activity. Enzyme replacement therapy (ERT) should be started before symptoms are apparent in order to achieve optimal outcomes. Early initiation of ERT in infantile-onset PD improves survival, reduces the need for ventilation, results in earlier independent walking, and enhances patient quality of life. Newborn screening (NBS) is the optimal approach for early diagnosis and treatment of PD. In NBS for PD, measurement of AαGlu enzyme activity in dried blood spots (DBSs) is conducted using fluorometry, tandem mass spectrometry, or digital microfluidic fluorometry. The presence of pseudodeficiency alleles, which are frequent in Asian populations, interferes with NBS for PD, and current NBS systems cannot discriminate between pseudodeficiency and cases with PD or potential PD. The combination of *GAA* gene analysis with NBS is essential for definitive diagnoses of PD. In this review, we introduce our experiences and discuss NBS programs for PD implemented in various countries.

Keywords: Pompe disease; newborn screening; pseudodeficiency; genotype-phenotype correlation; treatment and follow-up

---

1. Introduction

Glycogen storage disease type II (OMIM 232300), also known as Pompe disease (PD), is an autosomal recessive disorder caused by a defect in α-glucosidase (AαGlu; EC 3.2.1.20/3), resulting in the accumulation of lysosomal glycogen in skeletal and heart muscles [1]. The rates of accumulation and tissue damage depend on the residual enzyme activity. Patients with infantile-onset PD (IOPD) exhibit nearly complete absence of AαGlu activity and develop hypotonia and hypertrophic cardiomyopathy during infancy. Patients with IOPD eventually die of cardiorespiratory failure because massive amounts of glycogen accumulate in their skeletal and heart muscles. Patients with late-onset PD (LOPD) who exhibit marked reductions in AαGlu activity exhibit skeletal muscle dysfunction but rarely present with cardiac muscle disorders. The onset time and phenotypes of LOPD are variable, and patients are likely to exhibit manifestations in the fifth decade or later in life [2]. Enzyme replacement therapy (ERT) is essential for the treatment of IOPD [3,4]. ERT should be started before symptoms are clearly present, prior to the development of irreversible damage, to achieve optimal outcomes [5]. Early initiation of ERT can improve survival rates and quality of life in patients with IOPD, reducing the need for ventilation and leading to earlier independent walking [6].

Newborn screening (NBS) is an optimal approach for early diagnosis and treatment of IOPD. NBS, including pilot studies, is currently being carried out in several countries worldwide. Here, we describe NBS programs for PD, the diagnostic algorithm for PD, AαGlu enzyme assays using dried blood spots (DBSs) and fibroblasts, *GAA* gene analysis, and pseudodeficiency in *GAA*. Additionally, we discuss optimal treatments for PD, the current status of NBS worldwide, and future challenges in the development of NBS programs.

2. NBS Program for PD

NBS for PD is currently performed in Taiwan, Japan, and several states in the United States of America (USA). Current NBS systems measure AαGlu enzyme activity in DBSs.

In Japan, DBSs are prepared at maternity clinics or obstetrics departments using standard procedures at 4–6 days after birth for newborn mass screening according to public health system guidelines. After dropping blood spots onto filter papers (Toyo Roshi Kaisha, Ltd., Tokyo, Japan), DBSs are dried for at least 4 h at room temperature and sent to the Newborn Screening Center in Kumamoto within 1 week after preparation. The AαGlu activity in DBSs is then analyzed. The cutoff values in the AαGlu activity assay using DBSs differ for each research group but are set between 0.1 and 0.5 percentile values for the population or 20% to 30% of the mean value of the population. For newborns with values less than the cutoff values, a second AαGlu activity assay and *GAA* gene analysis are then performed [7].

At our institution, NBS for PD is performed in three steps (Figure 1). In the first step, newborns with AαGlu activity under the cutoff value of 6.5 pmol/h/disk (10% of the median value in the population) are recalled, and their DBSs are evaluated again. In the second step, using the Ba/Zn method, newborns with AαGlu activity under the cutoff value of 5.5 pmol/h/disk are called to the hospital within 2 months for a clinical examination. The infants are examined using physical and biochemical assays to confirm symptomatic signs of IOPD, and a third AαGlu assay is also performed. Finally, *GAA* gene analysis is performed in newborns with AαGlu activity under the cutoff value of 4.0 pmol/h/disk. The period after birth until the result of the first AαGlu assay is acquired is 1–2 weeks, and the period until the result of the second AαGlu assay is acquired is within 4 weeks. Thereafter, the period from birth until clinical examination is within 2 months, and the period from birth until *GAA* gene analysis and final diagnosis is up to 6 months [7].

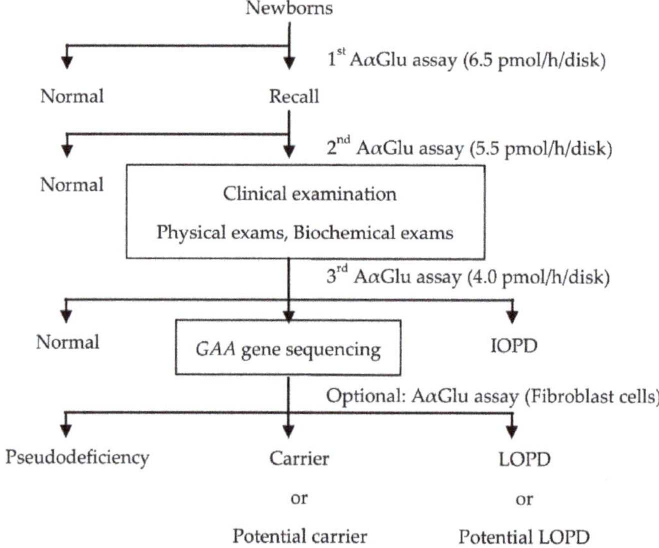

Figure 1. Flow chart of newborn screening (NBS) for Pompe disease (PD) in Japan. IOPD: infant-onset Pompe disease; LOPD: late-onset Pompe disease.

A definitive diagnosis of PD is achieved in patients harboring two known pathogenic *GAA* variants with decreased AαGlu activity in the blood (leukocytes, DBSs, isolated lymphocytes) or another tissue, such as fibroblast. A probable diagnosis for PD can be made in patients with decreased enzyme activity but ambiguous *GAA* gene analysis owing to the presence of molecular variants of unknown significance (VOUS). Moreover, the prevalence of pseudodeficiency alleles is high in Asian populations.

Figure 2 shows the diagnostic algorithm for PD. Clinicians can definitively diagnose patients with IOPD if they present with certain clinical manifestations, including heart or skeletal muscle deficiencies. Patients with LOPD definitively diagnosed by *GAA* gene analysis will need to be regularly followed up for the development of signs or symptoms related to PD, even if their *GAA* gene variants are known, because there is considerable variation in how and when patients will present symptoms. Patients with one or no known variants exhibiting decreased enzymatic activity should receive additional tests, including physical examinations, cardiac evaluations, AαGlu activity assays in fibroblasts, urinary glucotetrasaccharide (HEX4) and blood creatine kinase (CK) analyses, and/or parental DNA analyses. Through these additional tests, patients with one or no known variants may be diagnosed with LOPD, potential LOPD, or non-LOPD (carrier or pseudodeficiency) [8].

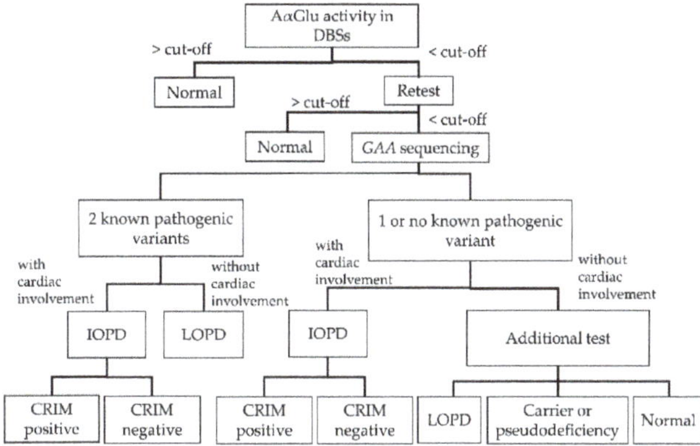

Figure 2. Flow chart of diagnosis for PD (modified from the Pompe Disease Newborn Screening Working Group [8]). DBS: dried blood spot; CRIM: cross-reactive immunological material.

3. AαGlu Enzyme Assay

In NBS for PD, AαGlu activity in DBSs is measured. Conventionally, the AαGlu activities in lymphocytes, fibroblasts, and skeletal muscles are analyzed for the diagnosis of PD [9]. Neutrophils in the blood contain maltase glucoamylase, another type of α-glucosidase. Because the pH at which this enzyme functions is consistent with that of AαGlu, the AαGlu activity assays in the blood are likely to result in false-negative results for defects in AαGlu [9]. However, large-scale NBS using DBSs has become possible owing to the use of acarbose, which inhibits the activity of maltase glucoamylase [10,11].

Measurement of AαGlu enzyme activity in DBSs can be carried out using fluorometry with the fluorogenic substrates of 4-methylumbelliferyl α-D-glucopyranoside (4MU-αGlc) [12], tandem mass spectrometry (MS/MS) [11], or digital microfluidic fluorometry [13]. Tandem mass spectrometry using mass-differentiated internal standards can quantify the corresponding enzymatic products and enables multiplex assays of a set of corresponding enzymes that cause lysosomal storage disorders (LSDs), such as PD, mucopolysaccharidosis (MPS), Fabry disease (FD), Gaucher disease (GD), Krabbe disease, and Niemann–Pick A/B disease. Additionally, digital microfluidics, a type of fluorometry, can be used to perform multiple assays of enzymes in the lysosome [14]. Tandem mass spectrometry is considered superior to fluorometry in terms of sensitivity, and some reports have shown that tandem mass spectrometry can distinguish pseudodeficiencies, which cannot be identified in patients with PD using fluorometry [15]. However, this is controversial because no reports have compared the outcomes of tandem mass spectrometry with those of fluorometry in the same samples. Moreover, the cutoff values used in NBS differ according to institution and region owing to factors such as differences in samples, measurement instruments, and humidity.

4. *GAA* Gene Analysis

GAA gene analysis is essential for definitive diagnosis of PD. The *GAA* gene is located on chromosome 17q25. It is 18.3 kb long, contains 20 exons, and encodes 952 amino acids. At present, more than 900 variants are registered in the ClinVar [16] or Pompe disease *GAA* variant databases [17], and the numbers are increasing. About two-thirds of these variants are classified by clinical significance and the other third of the variants are VOUS. Although bioinformatic tools such as Polyphen-2 [18], Human Splicing Finder [19], and Mutation Tester [20] are useful to estimate the pathogenicity of VOUS, these tools are insufficient for making the diagnosis. The progression of symptoms, treatment, and its outcomes are most important. Follow-up studies on the patients or potential patients with PD are essential and will help clinicians to diagnosis and determine proper treatment of patients with VOUS.

5. Pseudodeficiency

A pseudodeficiency allele is a change in the *GAA* gene sequence that results in AαGlu enzyme activity reduction, but is not enough to cause PD [21,22]. In our previous pilot program, the presence of pseudodeficiency alleles was shown to interfere with NBS for PD [23]. This suggests that NBS for PD must be able to distinguish PD cases and those with pseudodeficiency alleles in the *GAA* gene sequence. Asian patients are frequently homozygous or heterozygous for these pseudodeficiency alleles c.[1726G>A; 2965G>A] (p.[G576S; E689K]). Moreover, pseudodeficiency variants such as c. 1726G>A (p. G576S) are modifiers of pathogenic variants, which can result in greater reductions in GAA enzyme activity than with only the pathogenic variants [24]. Therefore, evaluation of pseudodeficiency in the *GAA* gene is essential in NBS.

Attempts have been made to distinguish cases with pseudodeficiency from patients with PD by methods other than *GAA* gene analysis; however, no reports have described a successful approach to achieve this goal.

6. AαGlu Activity in Fibroblasts

The measurement of AαGlu activity in fibroblasts was previously the gold standard for the diagnosis of PD [9]. This method can exclude the contribution of maltase glucoamylase activity. However, AαGlu assays in fibroblasts are difficult to perform as an NBS approach because acquiring fibroblasts for use in the AαGlu assay requires a skin biopsy, which is invasive and requires lengthy fibroblast culturing times. Despite this, AαGlu assays in fibroblasts are thought to be useful as additional tests for definitive diagnosis, even if *GAA* gene analysis cannot be used for the diagnosis of PD due to the presence of VOUS or only one known pathogenic variant for PD.

7. Cross-Reactive Immunological Material (CRIM)

Patients with IOPD should receive early ERT immediately after diagnosis of PD. Moreover, patients should undergo evaluation of CRIM before receiving ERT [25]. Patients with IOPD exhibiting residual AαGlu enzyme activity are CRIM-positive, and patients with IOPD exhibiting no residual AαGlu enzyme activity are CRIM-negative. CRIM-negative patients develop neutralizing antibodies for recombinant human lysosomal α-glucosidase (rhGAA) when receiving ERT, which impairs the effects of ERT [25,26].

In previous approaches for the evaluation of CRIM, residual AαGlu enzyme activity in fibroblasts was measured using an invasive method that required a long time to obtain results. Currently, the outcomes of *GAA* gene analysis contribute to estimations of the outcomes of CRIM [27].

8. ERT

ERT for patients with IOPD should be initiated as early as possible before irreversible damage occurs. Yang et al. indicated that early identification of patients with IOPD allows for very early initiation of ERT. Starting ERT even a few days earlier may lead to better patient outcomes [28]. Starting ERT early is effective in patients with LOPD as well; however, early ERT before presentation of signs or symptoms in patients with LOPD is generally not recommended [29]. Administration of ERT in the absence of symptoms of LOPD, even when blood CK and urine HEX4 are elevated, is avoided. The recommendation of ERT for patients with LOPD remains controversial.

Patients with LOPD should receive regular follow-ups, and levels of markers, such as blood CK and urine HEX4, should be monitored [30]. If patients with LOPD presenting symptoms of PD did not undergo NBS, their diagnosis and treatment are often delayed [31]. However, patients with LOPD diagnosed using NBS can be followed up and receive therapy immediately after presentation of the symptoms of PD [32]; treatment of these patients should not be delayed.

9. Immunomodulation to ERT

A variety of immunomodulation therapies have been developed to prevent the generation of neutralizing antibodies to rhGAA that would impair the effect of ERT. The immunomodulation therapies are elimination therapy for the neutralizing antibodies to rhGAA that have already been generated or prevention therapy for avoiding generation of the neutralizing antibodies before ERT. Rituximab, methotrexate, and intravenous immunoglobulins are often used for the immunomodulation therapies.

Prevention therapy has higher cost benefit than elimination therapy. Therefore, it is beneficial to identify CRIM-negative patients with IOPD before ERT in order to prevent generating neutralizing antibodies. Even some CRIM-positive IOPD patients are likely to develop neutralizing antibodies. The individualized T cell epitope measure scoring method, using a combination of individualized Human Leukocyte Antigen (HLA)-binding predictions and GAA genotype, may predict patient-specific risk of developing neutralizing antibodies to rhGAA [33].

Most patients with LOPD develop IgG antibodies to rhGAA, typically within 3 months of initiation of treatment [4]. Moreover, some patients who develop high, sustained antibody titers may have poorer clinical responses to treatment. Patients with IOPD or LOPD receiving ERT should routinely undergo tests for neutralizing antibodies for rhGAA. In rare cases in which high neutralizing antibodies interfere with the effects of ERT in patients with PD, we should consider the administration of immunosuppressants with ERT as well as discontinuing ERT.

10. The Follow-Up Period

Patients with IOPD who were diagnosed by NBS and received ERT should undergo regular follow-ups to assess treatment efficacy, onset of new symptoms, and deterioration of symptoms every month for the first 6 months or more [9,30]. In particular, cardiac evaluation is essential every month in the first 4 months of life and every 1–2 months thereafter. Clinicians should measure anti-rhGAA antibodies regardless of the state of CRIM. Patients with IOPD require immunosuppressants when they develop high titers of anti-rhGAA antibodies [34].

Patients with asymptomatic LOPD should undergo regular follow-up every 3 months during the first year after diagnosis. If they remain free of symptoms for 12 months, follow-ups every 3–12 months is required. Patients should receive ERT if they present signs or symptoms of PD.

Patients with symptomatic LOPD receiving ERT should undergo regular follow-ups monthly for 4 months after receiving ERT and then every 3 months thereafter, including monitoring for antibodies [30]. Because blood CK, aspartate aminotransferase (AST), alanine aminotransferase (ALT) levels, and urine HEX4 levels may increase before the onset of PD symptoms, these markers should be assessed regularly.

11. NBS Programs for PD Worldwide

The Newborn Screening Center at the National Taiwan University Hospital initiated an NBS program for PD in 2005. The outcomes of this large-scale NBS for PD in Taiwan demonstrated that the survival rates and ventilation-free rates of patients who were diagnosed with IOPD by NBS and received early ERT were higher than those of patients with IOPD who received ERT after presenting symptoms for IOPD [6,24]. Several regions in Japan also started NBS for PD in April, 2013 [7]. Moreover, the USA added PD to the Recommended Uniform Screening Panel (RUSP) and started NBS for PD in 2015. Currently, several countries worldwide have started pilot or regular NBS programs for PD (Table 1). The number and frequency of pseudodeficiencies and carriers between Japan and Taiwan are shown in Table 2 [35]; those in each country are displayed in Table 3.

Table 1. Summary of NBS programs for Pompe disease.

Country (No. of Newborns)	No. of Recalls (%)	No. of Patients (Prevalence) IOPD	No. of Patients (Prevalence) LOPD	Screening Method	Frequently Detected Pathogenic Variants	Reference
Taiwan (473,738)	2210 (0.47)	9 (1/52,638)	19 (1/24,934)	fluorometric assay	c.1935C>A, c.2238G>C	Chiang et al. (2012) [36]
Taiwan (191,786)	874 (0.46)	5 (1/38,357)	11 (1/17,435)	MS/MS	c.1935C>A, c.[752C>T; 761C>T]	Liao et al. (2014) [37]
Taiwan (64,148)	92 (0.14)	1 (1/64,148)	5 (1/12,830)	MS/MS	NA	Chiang et al. (2018) [38]
Japan (103,204)	225 (0.24)	0	3 (1/34,401)	fluorometric assay	c.[752C>T; 761C>T], c.317G>A	Momosaki et al. (2019) [7]
Washington (111,554)	17 (0.02)	0	4 (1/27,889)	MS/MS	c.-32-13T>G	Scott et al. (2013) [39]
Washington (44,047)	9 (0.02)	0	1 (1/44,047)	MS/MS	c.2168del13ins10	Elliott et al. (2016) [40]
Missouri (43,701)	18 (0.04)	3 (1/14,567)	5 (1/8740)	DMF	NA	Hopkins et al. (2015) [41]
Illinois (219,973)	139 (0.06)	2 (1/149,987)	8 (1/37,497)	MS/MS	c.-32-13T>G	Burton et al. (2017) [42]
New York (18,105)	6 (0.03)	0	1 (1/18,105)	MS/MS	c.-32-13T>G	Wasserstein et al. (2019) [43]
Austria (34,736)	5 (0.01)	0	4 (1/8684)	MS/MS	c.896T>C, c.-32-13T>G	Mechtler et al. (2012) [44]
Hungary (40,024)	163 (0.41)	0	9 (1/4447)	MS/MS	c.664G>A, c.-32-13T>G	Wittmann et al. (2012) [45]
Italy (44,411)	8 (0.02)	2 (1/22,206)	0	MS/MS	c.-32-13T>G, c.236_246del	Burlina et al. (2018) [46]
Mexico (20,018)	19 (0.09)	0	1 (1/ 20,018)	MS/MS	c.1375G>A	Navarrete-Martínez et al. (2017) [47]
Brazil (103,204)	NA	0	0	DMF	-	Bravo et al. (2017) [48]

USA is indicated as a bracket label grouping the Washington, Missouri, Illinois, New York rows.

NA: not available.

Table 2. The distribution of pseudodeficiency alleles and PD-associated variants in newborns who were detected by NBS for PD.

Country/ Reference	Pseudodeficiency Alleles	No. of PD-Associated Variants 0	No. of PD-Associated Variants 1	No. of PD-Associated Variants 2	Prevalence (%)
Japan (*n* = 103,204)/ Momosaki, et al. (2019) [7]	Homozygous	24	8	0	32/71 (45.1%)
	Heterozygous	0	35	3	38/71 (53.5%)
	None	0	0	1	1/71 (1.5%)
	Diagnosis	Pseudodeficiency 24/71 (33.8%)	Carrier or potential carrier 43/71 (60.6%)	Patient or potential patient 4/71 (5.6%)	
Taiwan (*n* = 132,538)/ Labrousse et al. (2009) [35]	Homozygous	36	32	0	68/104 (65.4%)
	Heterozygous	0	27	7	34/104 (32.7%)
	None	0	0	2	2/104 (1.9%)
	Diagnosis	Pseudodeficiency 36/104 (34.6%)	Carrier or potential carrier 59/104 (56.7%)	Patient or potential patient 9/104 (8.7%)	

Table 3. Number of pseudodeficiencies and carriers in each country.

Country (No. of Newborns)		Pseudodeficiency (with 0 PD-Associated Variants)		Carrier or Potential Carrier (with 1 PD-Associated Variant)		Reference
		No.	Genotype	No.	Genotype	
USA	Washington (111,554)	6	pseudodeficiency allele/wt. ($n = 6$)	7	pathogenic allele/wt. ($n = 4$) pathogenic allele/pseudodeficiency allele ($n = 3$)	Scott et al. (2013) [39]
	Washington (44,047)	0		0		Elliott et al. (2016) [40]
	Missouri (43,701)	2	NA	3	NA	Hopkins et al. (2015) [41]
	Illinois (219,973)	15	NA	19	NA	Burton et al. (2017) [42]
	New York (18,105)	3	c.[1726G>A; 2065G>A]/c.[1726G>A; 2065G>A] ($n = 2$) c.[1726G>A; 2065G>A]/wt.	2	c.2560C>T(VOUS)/c.858+20_858+26del*(Predicted Benign) c.[1726G>A; 2065G>A]/c.-32-13T>G	Wasserstein et al. (2019) [43]
Austria (34,736)		0		0		Mechtler et al. (2012) [44]
Hungary (40,024)		0		17	NA	Wittmann et al. (2012) [45]
Italy (44,411)		0		2	c.[1726G>A; 2065G>A]/c.-32-13T>G c.1726G>A/c.-32-13T>G	Burlina et al. (2018) [46]
Mexico (20,018)		8	c.[1726G>A; 2065G>A]/c.[1726G>A; 2065G>A] ($n = 6$) c.[1726G>A;2065G>A]/wt. ($n = 2$)	2	c.[1726G>A; 2065G>A]/c.-32-13T>G ($n = 2$)	Navarrete-Martínez et al. (2017) [47]
Brazil (103,204)		0		1	c.[1726G>A; 2065G>A]/c.-32-13T>G	Bravo et al. (2017) [48]

NA: not available.

11.1. Taiwan

Chien et al. performed the first large pilot NBS program to detect PD in newborns in Taiwan using a fluorometric enzymatic assay to determine AαGlu activity in DBSs. They conducted a pilot NBS of 132,538 newborns, accounting for almost 45% of newborns in Taiwan, between October 2005 and March 2007. Of the 132,538 newborns screened, 1093 (0.82%) underwent repeated DBS sampling, and 121 (0.091%) newborns were recalled for additional evaluation. PD was identified in 4 newborns (3 IOPD and 1 LOPD) [49]. Owing to this outcome, NBS for PD is now regularly conducted in Taiwan. Moreover, they identified 9 patients with IOPD and 19 patients with LOPD among 473,738 newborns by NBS for PD between October 2005 and December 2011 [36]. They launched a four-plex MS/MS LSD newborn screening test also including AαGlu (PD), acid α-galactosidase (FD), acid α-glucocerebrosidase (GD), and acid α-L-iduronidase (MPSI) in 2015. Through 2017, 64,148 newborns were screened for these four LSDs using their system. The cutoff levels in this new NBS system were established as 0.1 percentile of the population, or 13–15% of the normal mean. This NBS detected 20 infants with less than the cutoff value, and 1 patient with IOPD, 5 patients with LOPD, and 14 infants with false-positive results were identified [38].

Liao et al. reported the results of 191,786 newborns evaluated in an NBS program for PD using a system that could detect multiple LSDs by MS/MS from February 2010 to January 2013. After the initial DBS screening, 9 newborns were referred to hospitals directly with AαGlu values lower than the critical cutoff value (0.20 μmol/L/h) or combined with some clinical symptoms. In total, 874 (0.46%) newborns were recalled for second DBSs, 225 (0.12%) suspected newborns with decreased AαGlu activity were referred to hospitals, and 16 newborns were confirmed to have PD. In *GAA* gene analysis, 5 newborns were classified as IOPD and 11 newborns as LOPD. *GAA* gene analysis demonstrated that c.1935C>A (p.D645E) was detected in all cases of IOPD, c.[752C>T; 761C>T] (p.[S251L; S254L]) was detected in 8 cases of LOPD, and the c.1726G>A (p.G576S) pseudodeficiency variant was detected in 2 cases of IOPD and 5 cases of LOPD. The variants c.1840A>G (p.614A), c.2647-23delT, c.1054C>T (p.Q352*), IVS7+2T>C, and IVS17-5T>C were identified as novel variants. The false-positive rate in the tandem mass method was similar to that in fluorometric assays [37].

11.2. Japan

We started a pilot study of NBS for FD using 4-methylumbelliferyl-α-d-galactopyranoside (4MU-αGlc) in August 2006 and have conducted NBS of 5 LSDs, including PD, FD, GD, MPSI, and MPSII [50,51].

We reported the results of NBS for PD using 4MU-αGlc in 103,204 newborns [7]. Among these newborns, 225 newborns were retested using a second AαGlu assay, and 111 newborns with low AαGlu activities under the cutoff in the second AαGlu assay were evaluated for IOPD detection using physical and biochemical examinations (CK, ALT, AST, and lactate dehydrogenase), echocardiogram assessments, and a third AαGlu assay. For the 71 newborns with low AαGlu activity under the cutoff in the third AαGlu assay, *GAA* gene sequencing was performed using NGS. The AαGlu activities in fibroblasts were measured in 32 of the 71 newborns. In this study, no newborns developed IOPD, and 50 variants were detected. Eight variants were novel: c.547-67C>G, c.692+38C>T, c.1082C>A (p.P361Q), c.1244C>T (p.T415M), c.1552-52C>A, c.1638+43G>T, c.2003A>G (p.Y668C), and c.2055C>G (p.Y685*). The most common mutation was c.[752C>T;761C>T] (p.[S251L; S254L]), accounting for 20 alleles (14.1%, 20/142). The pseudodeficiency alleles c.1726G>A (p.G576S) and c.2065G>A (p.E689K) were detected in 71.8% (102/142) and 72.5% (103/142) of all newborns with low AαGlu activity, respectively.

This study identified 3 newborns with potential LOPD but without IOPD detection. Although these 3 patients developed no symptoms related to PD and received no treatment, the c.317G>A (p.R106H), c.1244C>T (p.T415M), and c.2003A>G (p.Y668C) mutations in these 3 patients were considered novel mutations. The prevalence of potential PD was 1 per 34,401 births. Newborns with both pseudodeficiency alleles and PD-associated pathogenic variants were detected in Japan as well as in Taiwan (Table 2). Table S1 displays the distribution of mutations and predictably pathogenic variants in NBS for PD in Japan.

11.3. USA

In 2008, the Advisory Committee on Heritable Disorders in Newborns and Children (ACHDNC) evaluated the NBS system for PD. The committee found significant evidence gaps related to the accuracy of screening and to the benefits and harms of presymptomatic diagnosis and precluded recommendation of NBS of PD for the Recommended Uniform Screening Panel (RUSP). In 2013, the ACHDNC reconsidered PD after it was nominated again. Based in part on new information presented to the ACHDNC by the external condition review workgroup, NBS for PD was recommended for addition to the RUSP and was added in March 2015. Prosser et al. estimated that screening 4 million babies born each year in the United States would detect 134 cases with PD including 40 cases with IOPD, compared with 36 cases detected clinically without screening [52]. NBS would also identify 94 cases of LOPD that might not become symptomatic for decades. By identifying 40 cases with IOPD, NBS would avert 13 deaths and identify 26 individuals requiring mechanical ventilation by the age of 36 months.

11.3.1. Washington

In 2013, Scott et al. reported screening results for more than 110,000 newborns in Washington. They detected PD, FD, GD, MPSI, MPSII (α-L-iduronide-2-sulfatase), Niemann–Pick A/B disease (acid sphingomyelinase), MPSIV-A (galactose-6-sulfate sulfatase), MPSVI (N-acetylgalactosamine-4-sulfatase), and Krabbe disease (galactocerebrosidase) using a technology which simultaneously measured multiple enzyme activities by MS/MS [39]. AαGlu activity in DBSs was measured in 111,544 cases. A cutoff value was established as 15% of the mean value. Seventeen samples had enzyme activities with less than the cutoff value. Four cases (2 cases with the homozygous IVS1-13T>G variant, 1 case with the compound heterozygous c.365T>A/c.1925T>A variant, and 1 case with the compound heterozygous IVS1-13T>G/c.1-17C>T variant) were confirmed to be patients with LOPD or potential LOPD, 4 cases had a single nucleotide change on one allele (carrier of PD), 3 cases were identified as carriers with an additional pseudodeficiency allele, and 6 cases were heterozygotes for a pseudodeficiency allele only. The prevalence of infants with LOPD or potential LOPD was 1 per 27,800 births.

Elliott et al. evaluated 43,000 newborns in NBS using a multiplex MS/MS enzymatic activity assay of 6 lysosomal enzymes for PD, FD, GD MPSI, Niemann–Pick A/B disease, and Krabbe disease. The cutoff value was established as 10% of the mean value. A newborn with p.G576S/p.T602I (a probable low activity variant) and a newborn with the homozygous c.2168del13ins10 (p.A724Gfs*44) variant (a frameshift variant leading to a nearby stop codon) were identified [40].

11.3.2. Missouri

A full-population pilot study of 43,701 newborns using DBSs in a multiplex fluorometric enzymatic assay for detecting PD, FD, GD, and MPSI was performed in Missouri on January 11, 2013 [41]. The cutoff values for AαGlu activities were set at the 0.17 percentile in the pilot study. Of the 18 cases that screened positive for AαGlu deficiency, 3 were diagnosed with IOPD, 3 were classified as LOPD, 2 were classified as potential LOPD or VOUS, 2 had pseudodeficiencies, and 3 were carriers. The prevalence of PD was 1 per 8740 births.

11.3.3. Illinois

The Newborn Screening Laboratory of Illinois performed NBS for 5 LSD-associated enzymes, including PD, FD, MPSI, and Niemann–Pick A/B disease, among 219,973 newborn DBSs using MS/MS from November 1, 2014 to August 31, 2016 [42]. In total, 139 (0.06%) had a positive or borderline test result in PD, necessitating additional testing. The cutoff values for AαGlu activities were defined as follows: positive = less than or equal to 18% of the daily median value, and borderline = greater than 18% and less than or equal to 22% of the daily median value. Ten cases of PD (two cases of IOPD and eight cases of LOPD) were detected. The frequency of PD was 1 per 21,997 births. Two infants diagnosed with IOPD developed elevated CK levels and clear evidence of cardiomyopathy at the time of initial evaluation which included chest radiography, electrocardiography, and echocardiography. These patients are regularly receiving ERT. Eight infants diagnosed with LOPD had either homozygous or compound heterozygous variants of the common splicing mutation c.-32-13T>G observed in patients with LOPD. *GAA* pseudodeficiency was detected in 15 infants, including 14 identified as Asian (4 Chinese, 3 Filipino, 2 Korean, 2 Indian, 1 Japanese, and 2 others). There were four infants with an undetermined classification or "potential PD."

11.3.4. New York

A pilot NBS study for 18,105 newborns was performed through October 1, 2014 [43]. Six cases were positive in the screen (1 case of LOPD, 3 cases of pseudodeficiency, and 2 carriers) with a mean AαGlu activity of less than or equal to 15% of the daily mean activity, yielding a referral rate of 0.033%. One case with LOPD had the homozygous c.-32-13T>G variant and low leukocyte AαGlu activity; however, examination results and laboratory values were normal and HEX4 testing was negative. Two cases were homozygous for the common pseudodeficiency allele, and another case carried one copy of the pseudodeficiency allele; leukocyte AαGlu activity in these infants ranged from low to normal. One case was found to have c.2560C>T (p.R854*) and an intronic variant believed to be benign. This infant was diagnosed as being a carrier because AαGlu activity was high.

11.4. Austria

DBSs of 34,736 newborns were collected consecutively in the national routine Austrian NBS program from January 2010 to July 2010 and analyzed for enzyme activities of acid β-glucocerebrosidase, α-galactosidase, α-glucosidase, and acid sphingomyelinase by electrospray ionization MS/MS [44]. The cutoff value for AαGlu was based on the 0.1 percentile value from 5000 cases of AαGlu activity. This first-line screening for low AαGlu activity detected 25 cases, and retests showed 5 cases with low

AαGlu activities. Sequence analyses of the *GAA* gene identified 4 cases with PD presenting homozygous c.896T>C (p.L299P), homozygous c.-32-13T>G, or compound heterozygous c.-32-13T>G/c.1551+1G>A (V480_I517del) variants. The prevalence of PD was 1 per 8684 births.

11.5. Hungary

In the Hungarian NBS program, 40,024 newborns were screened for PD, FD, GD, and Niemann–Pick A/B disease using MS/MS [45]. The 0.25th–0.5th percentile of AαGlu activities from 1000 cases were defined as the cutoff values; 663 cases (1.66%) were submitted for retesting. Among them, 163 cases (0.41%) had abnormal AαGlu activities in DBSs. After retesting, *GAA* gene analysis was performed in 64 (0.160%) cases with low AαGlu activity. *GAA* gene analysis detected 9 cases with PD and 25 carriers for PD. Three cases remained uncertain owing to inclusive *GAA* variants. Five cases had c.664G>A (p.V222M)/c.664G>A (p.V222M), and the other cases had c.-32-13T>G/c.-32-13T>G, c.664G>A (p.V222M)/c.2174G>A (p.R725Q), c.1216G>A (p.D406N)/c.1409A>C (p.N470T), and c.1552-3C>G/c.1552-3C>G. The 25 carriers had *GAA* gene variants related to PD, including -32-13T>G, c.307T>G (p.C103G), c.664G>A (p.V222M), c.763C>T (p.Q255X), c.841C>T (p.R281W), c.875A>G, c.1437+1G>A, c.1468 T>C (p.F490L), c.1552-3C>G, c.1903A>G (p.N635D), c.2237G>T (p.W746L), and c.2482-2A>G. The prevalence of PD was 1 per 20,012 births.

11.6. Italy

NBS programs for PD, FD, GD, and MPSI were performed using DBSs from 44,411 newborns by multiplexed MS/MS with the NeoLSD assay system in northeastern Italy from September 2015 to January 2017 [46]. Among the 44,411 newborns screened for the four LSDs, 40 cases (0.09%) had enzyme activity below 0.2 multiples of the median and were recalled for collection of a second DBS. Eight cases showed low AαGlu activities. Five patients underwent *GAA* gene analysis, and 2 cases were identified as juvenile types of PD because of the presence of the compound heterozygote of c.-32-13T>G (IVS1-13T>G)/c.236_246del (p.P79Rfs*12); elevated blood CK, AST, and ALT levels; and slightly enlarged heart findings. These patients underwent ERT immediately after diagnosis of PD. One case was classified as VOUS, and 2 cases were carriers for *GAA* variants. The frequency of PD was 1 per 22,205 births.

A total of 3403 newborns (1702 males and 1701 females) in the Umbria area of central Italy were screened for PD, FD, GD, and MPSI using DBSs [53]. The cutoff value was established as 35% of the normal median AαGlu activity. Although 3 cases showed AαGlu activities with less than the cutoff value in DBSs, none of these cases showed abnormal lymphocyte AαGlu activities. This pilot study identified no infants with PD.

11.7. Mexico

NBS for 6 LSDs, including PD, FD, GD MPSI, Niemann–Pick A/B disease, and Krabbe disease, using a multiplex MS/MS enzymatic assay was performed in 20,018 newborns (10,241 males and 9777 females) from July 1, 2012 to April 30, 2016 [47]. Nineteen presumptive positive infants showed low AαGlu activity in the first DBS. In the second AαGlu activity test, 16 infants were positive. Among these infants, 11 showed low leukocyte AαGlu activity. Five infants could not complete the protocol, three of whom lost healthcare insurance and 2 of whom refused to continue the protocol. In 11 infants with low leukocyte AαGlu activity, 10 infants harbored a pseudodeficiency variant associated with low AαGlu enzyme activity, but without signs of PD (homozygous or heterozygous for c.[1726G>A; 2065G>A] (p.[G576S; E689K])). Moreover, 2 infants harbored a compound heterozygous variant for the pseudodeficiency allele and the -32-13 T> G variant. The patient with potential LOPD presented with the c.1375G>A (p.D459N) variant, which was reported as an LOPD variant, and the VOUS c.1220A>G (p.Y407C) variant, which was predicted to be pathogenic.

11.8. Brazil

A pilot NBS study used a digital microfluidic platform to simultaneously measure the activities of AαGlu, acid β-glucosidase, α-galactosidase, and iduronidase to screen for PD, GD, FD, and MPSI in DBSs, respectively [54]. The cutoff value for AαGlu was defined as less than 30% of the mean enzyme activity in samples from 1000 unaffected babies. One case was identified to be pseudodeficiency, and no infants were identified with PD [48,55].

12. Genotype–Phenotype Correlation

Known variants in PD are registered in PD variant databases, including the Pompe Center (http://www.pompevariantdatabase.nl/pompe_mutations_list.php?orderby=aMut_ID1) and ClinVar (http://www.ncbi.nlm.nih.gov/clinvar). Moreover, in PolyPhen-2 (http://genetics.bwh.harvard.edu/pph2), the impact of missense variants on amino acid substitutions is evaluated. Human Splicing Finder (http://www.umd.be/HSF3/) can predict the impact of splicing abnormalities.

c.525delT (p.E176Rfs*45), a common frameshift variant in the Netherlands [56], c.1935C>A (p.D645E), a common missense variant in Taiwan [6], and c.2560C>T (p.R854*), a common nonsense variant in Africa [57], are considered variants for IOPD. c.-32-13T>G is commonly detected in Caucasian patients with LOPD, and patients with the homozygous c.-32-13T>G variant rarely develop cardiac manifestations as infants [58]. c.-32-13T>G is considered a variant in LOPD. However, because even patients with the same variants can develop both IOPD and LOPD phenotypes, the phenotype cannot be predicted from gene analysis alone. Moreover, the distribution of gene variants differs in each region. In the future, the accumulation of genetic information for patients in each region will be essential for predicting phenotypes.

13. Potential Concern of Screening for PD

NBS not only detects patients with IOPD but also patients with LOPD and potential LOPD. The effectiveness of NBS for identification of patients with LOPD and potential LOPD is controversial. Patients with LOPD who would present with PD symptoms within 2–3 years can receive medication before the progression of the symptoms, due to early diagnosis of LOPD through NBS. However, the demerits of NBS for patients with LOPD and potential LOPD should be considered because the time from diagnosis to presentation of PD symptoms may be more than 10 years, and some patients with LOPD or potential LOPD may not ever develop PD symptoms. Therefore, problems such as the psychological stress for the family due to LOPD and potential LOPD diagnosis [59], the cost and time of visiting hospitals and receiving medical examinations, and the potential of receiving overtreatment are issues likely to occur.

As shown in the guidelines for PD promulgated by the Pompe Disease Newborn Screening Working Group [30], clinicians should consider the need for psychosocial support for families during follow-ups for presymptomatic patients with LOPD. As more patients or potential patients with LOPD are diagnosed and followed up, it has been demonstrated that LOPD causes more symptoms than proximal myopathy or respiratory failure; it is a multiorgan disorder involving muscular, respiratory, musculoskeletal, peripheral nervous, vascular, cardiac, and gastrointestinal systems. Common symptoms reported in LOPD are proximal muscle weakness, trunk muscle weakness, exercise intolerance, shortness of breath, impaired cough, and gait difficulties. Because LOPD is a multisystemic disease, clinicians should be aware of all known symptoms and indicators in order to prevent delayed diagnoses and misdiagnoses. The understanding of the natural history of LOPD is advanced in the use of ERT. In the future, the disease concept of LOPD as well as IOPD will be more established.

14. Future Challenges

NBS programs for PD can contribute to early detection and early intervention in patients with IOPD and LOPD. Early ERT can change the natural clinical course and result in better outcomes in patients with IOPD. Because of changes in the natural clinical course, neurological manifestations, which had not previously been discussed, have become apparent [60]. For example, some patients with IOPD receiving ERT present with learning disorders as neurological manifestations. Currently available rhGAA therapy cannot cross the blood–brain barrier [61,62]. Moreover, patients with LOPD detected in NBS can receive follow-up and early intervention before exhibiting deterioration of PD symptoms [32].

Nevertheless, currently available NBS programs that evaluate AαGlu activity in DBSs, even those using tandem mass spectrometry, cannot discriminate pseudodeficiency from cases of PD or potential PD. Because families of newborns with pseudodeficiency have anxiety related to the results of NBS, new NBS programs, such as those using a combination of AαGlu enzyme assays and *GAA* gene analysis from DBS, can be used to distinguish pseudodeficiency from PD or VOUS. Such approaches are urgently needed.

In the future, next-generation treatments, including chaperon therapy [63] and gene therapy for PD [64,65], are expected to have favorable outcomes. Therefore, many researchers should contribute to the development of novel, improved NBS programs and the spread of such NBS programs to more regions around the world.

Supplementary Materials: Supplementary materials can be found at http://www.mdpi.com/2409-515X/6/2/31/s1. Table S1: The distribution of mutations and predictably pathogenic variants in NBS for Pompe disease in Japan.

Author Contributions: Conceptualization, T.S., J.K. and K.N.; writing—original draft preparation, T.S. and J.K.; writing—review and editing, J.K. and K.N.; visualization, T.S. and J.K.; supervision, J.K. and K.N.; project administration, J.K. and K.N.; funding acquisition, K.N. All authors have read and agreed to the published version of the manuscript.

Funding: This review article was supported in part by a Grant-in-Aid for Research on Rare and Intractable Diseases, Health and Labor Sciences Research; a Grant-in-Aid for Pediatric Research from the Ministry of Health, Labor and Welfare; a Grant-in-Aid from the Japan Agency for Medical Research and Development; and a Grant-in-Aid for Scientific Research from the Ministry of Education, Culture, Sports, Science, and Technology.

Acknowledgments: We are grateful to Fumiko Nozaki, Naomi Yano, Ayuko Tateishi, Emi Harakawa, Yasuyo Sakamoto, Hiroko Nasu, and Matsumi Harada for their technical support in the NBS for PD in Japan. Special thanks for Keishin Sugawara for his expert advice.

Conflicts of Interest: The authors declare no conflict of interest.

References

1. Martiniuk, F.; Mehler, M.; Pellicer, A.; Tzall, S.; La Badie, G.; Hobart, C.; Ellenbogen, A.; Hirschhorn, R. Isolation of a cDNA for human acid α-glucosidase and detection of genetic heterogeneity for mRNA in three α-glucosidase-deficient patients. *Proc. Natl. Acad. Sci. USA* **1986**, *83*, 9641–9644. [CrossRef] [PubMed]

2. Hagemans, M.L.C.; Winkel, L.P.F.; Hop, W.C.J.; Reuser, A.J.J.; Van Doorn, P.A.; van der Ploeg, A.T. Disease severity in children and adults with Pompe disease related to age and disease duration. *Neurology* **2005**, *64*, 2139–2141. [CrossRef] [PubMed]

3. Kishnani, P.S.; Corzo, D.; Nicolino, M.; Byrne, B.; Mandel, H.; Hwu, W.L.; Leslie, N.; Levine, J.; Spencer, C.; McDonald, M.; et al. Recombinant human acid α-glucosidase: Major clinical benefits in infantile-onset Pompe disease. *Neurology* **2007**, *68*, 99–109. [CrossRef] [PubMed]

4. van der Ploeg, A.T.; Clemens, P.R.; Corzo, D.; Escolar, D.M.; Florence, J.; Groeneveld, G.J.; Herson, S.; Kishnani, P.S.; Laforet, P.; Lake, S.L.; et al. A Randomized Study of Alglucosidase Alfa in Late-Onset Pompe's Disease. *N. Engl. J. Med.* **2010**, *362*, 1396–1406. [CrossRef]

5. Kishnani, P.S.; Corzo, D.; Leslie, N.D.; Gruskin, D.; Van Der Ploeg, A.; Clancy, J.P.; Parini, R.; Morin, G.; Beck, M.; Bauer, M.S.; et al. Early treatment with alglucosidase alfa prolongs long-term survival of infants with pompe disease. *Pediatr. Res.* **2009**, *66*, 329–335. [CrossRef]

6. Chien, Y.-H.; Lee, N.-C.; Thurberg, B.L.; Chiang, S.-C.; Zhang, X.K.; Keutzer, J.; Huang, A.-C.; Wu, M.-H.; Huang, P.-H.; Tsai, F.-J.; et al. Pompe disease in infants: Improving the prognosis by newborn screening and early treatment. *Pediatrics* **2009**, *124*, e1116–e1125. [CrossRef]

7. Momosaki, K.; Kido, J.; Yoshida, S.; Sugawara, K.; Miyamoto, T.; Inoue, T.; Okumiya, T.; Matsumoto, S.; Endo, F.; Hirose, S.; et al. Newborn screening for Pompe disease in Japan: Report and literature review of mutations in the *GAA* gene in Japanese and Asian patients. *J. Hum. Genet.* **2019**, *64*, 741–755. [CrossRef]

8. Burton, B.K.; Kronn, D.F.; Hwu, W.-L.; Kishnani, P.S.; Pompe Disease Newborn Screening Working Group. The Initial Evaluation of Patients after Positive Newborn Screening: Recommended Algorithms Leading to a Confirmed Diagnosis of Pompe Disease. *Pediatrics* **2017**, *140*, S14–S23. [CrossRef]

9. Kishnani, P.S.; Steiner, R.D.; Bali, D.; Berger, K.; Byrne, B.J.; Case, L.E.; Crowley, J.F.; Downs, S.; Howell, R.R.; Kravitz, R.M.; et al. Pompe disease diagnosis and management guideline. *Genet. Med.* **2006**, *8*, 267–288. [CrossRef]

10. Jack, R.M.; Gordon, C.; Scott, C.R.; Kishnani, P.S.; Bali, D. The use of acarbose inhibition in the measurement of acid alpha-glucosidase activity in blood lymphocytes for the diagnosis of Pompe disease. *Genet. Med.* **2006**, *8*, 307–312. [CrossRef]

11. Li, Y.; Scott, C.R.; Chamoles, N.A.; Ghavami, A.; Pinto, B.M.; Turecek, F.; Gelb, M.H. Direct multiplex assay of lysosomal enzymes in dried blood spots for newborn screening. *Clin. Chem.* **2004**, *50*, 1785–1796. [CrossRef] [PubMed]

12. Chamoles, N.A.; Niizawa, G.; Blanco, M.; Gaggioli, D.; Casentini, C. Glycogen storage disease type II: Enzymatic screening in dried blood spots on filter paper. *Clin. Chim. Acta* **2004**, *347*, 97–102. [CrossRef] [PubMed]

13. Sista, R.S.; Wang, T.; Wu, N.; Graham, C.; Eckhardt, A.; Winger, T.; Srinivasan, V.; Bali, D.; Millington, D.S.; Pamula, V.K. Multiplex newborn screening for Pompe, Fabry, Hunter, Gaucher, and Hurler diseases using a digital microfluidic platform. *Clin. Chim. Acta* **2013**, *424*, 12–18. [CrossRef] [PubMed]

14. Graham, C.; Sista, R.S.; Kleinert, J.; Wu, N.; Eckhardt, A.; Bali, D.; Millington, D.S.; Pamula, V.K. Novel application of digital microfluidics for the detection of biotinidase deficiency in newborns. *Clin. Biochem.* **2013**, *46*, 1889–1891. [CrossRef]

15. Liao, H.-C.; Chan, M.-J.; Yang, C.-F.; Chiang, C.-C.; Niu, D.-M.; Huang, C.-K.; Gelb, M.H. Mass Spectrometry but Not Fluorimetry Distinguishes Affected and Pseudodeficiency Patients in Newborn Screening for Pompe Disease. *Clin. Chem.* **2017**, *63*, 1271–1277. [CrossRef]

16. Landrum, M.J.; Chitipiralla, S.; Brown, G.R.; Chen, C.; Gu, B.; Hart, J.; Hoffman, D.; Jang, W.; Kaur, K.; Liu, C.; et al. ClinVar: Improvements to accessing data. *Nucleic Acids Res.* **2020**, *48*, D835–D844. [CrossRef]

17. Niño, M.Y.; in 't Groen, S.L.M.; Bergsma, A.J.; Beek, N.A.M.E.; Kroos, M.; Hoogeveen-Westerveld, M.; Ploeg, A.T.; Pijnappel, W.W.M.P. Extension of the Pompe mutation database by linking disease-associated variants to clinical severity. *Hum. Mutat.* **2019**, *40*, 1954–1967. [CrossRef]

18. Adzhubei, I.; Jordan, D.M.; Sunyaev, S.R. Predicting Functional Effect of Human Missense Mutations Using PolyPhen-2. *Curr. Protoc. Hum. Genet.* **2013**. [CrossRef]

19. Desmet, F.-O.; Hamroun, D.; Lalande, M.; Collod-Béroud, G.; Claustres, M.; Béroud, C. Human Splicing Finder: An online bioinformatics tool to predict splicing signals. *Nucleic Acids Res.* **2009**, *37*, e67. [CrossRef]

20. Schwarz, J.M.; Cooper, D.N.; Schuelke, M.; Seelow, D. Mutationtaster2: Mutation prediction for the deep-sequencing age. *Nat. Methods* **2014**, *11*, 361–362. [CrossRef]

21. Kroos, M.A.; Mullaart, R.A.; Van Vliet, L.; Pomponio, R.J.; Amartino, H.; Kolodny, E.H.; Pastores, G.M.; Wevers, R.A.; Van der Ploeg, A.T.; Halley, D.J.J.J.; et al. p.[G576S.; E689K]: Pathogenic combination or polymorphism in Pompe disease? *Eur. J. Hum. Genet.* **2008**, *16*, 875–879. [CrossRef] [PubMed]

22. Tajima, Y.; Matsuzawa, F.; Aikawa, S.; Okumiya, T.; Yoshimizu, M.; Tsukimura, T.; Ikekita, M.; Tsujino, S.; Tsuji, A.; Edmunds, T.; et al. Structural and biochemical studies on Pompe disease and a "pseudodeficiency of acid α-glucosidase". *J. Hum. Genet.* **2007**, *52*, 898–906. [CrossRef] [PubMed]

23. Kumamoto, S.; Katafuchi, T.; Nakamura, K.; Endo, F.; Oda, E.; Okuyama, T.; Kroos, M.A.; Reuser, A.J.J.; Okumiya, T. High frequency of acid α-glucosidase pseudodeficiency complicates newborn screening for glycogen storage disease type II in the Japanese population. *Mol. Genet. Metab.* **2009**, *97*, 190–195. [CrossRef] [PubMed]

24. Chien, Y.-H.; Hwu, W.-L.; Lee, N.-C. Pompe Disease: Early Diagnosis and Early Treatment Make a Difference. *Pediatr. Neonatol.* **2013**, *54*, 219–227. [CrossRef]

25. Kishnani, P.S.; Goldenberg, P.C.; DeArmey, S.L.; Heller, J.; Benjamin, D.; Young, S.; Bali, D.; Smith, S.A.; Li, J.S.; Mandel, H.; et al. Cross-reactive immunologic material status affects treatment outcomes in Pompe disease infants. *Mol. Genet. Metab.* **2010**, *99*, 26–33. [CrossRef]

26. Banugaria, S.G.; Prater, S.N.; Ng, Y.K.; Kobori, J.A.; Finkel, R.S.; Ladda, R.L.; Chen, Y.T.; Rosenberg, A.S.; Kishnani, P.S. The impact of antibodies on clinical outcomes in diseases treated with therapeutic protein: Lessons learned from infantile Pompe disease. *Genet. Med.* **2011**, *13*, 729–736. [CrossRef]

27. Desai, A.K.; Kazi, Z.B.; Bali, D.S.; Kishnani, P.S. Characterization of immune response in Cross-Reactive Immunological Material (CRIM)-positive infantile Pompe disease patients treated with enzyme replacement therapy. *Mol. Genet. Metab. Rep.* **2019**, *20*, 100475. [CrossRef]

28. Yang, C.F.; Yang, C.C.; Liao, H.C.; Huang, L.Y.; Chiang, C.C.; Ho, H.C.; Lai, C.J.; Chu, T.H.; Yang, T.F.; Hsu, T.R.; et al. Very Early Treatment for Infantile-Onset Pompe Disease Contributes to Better Outcomes. *J. Pediatr.* **2016**, *169*, 174–180.e1. [CrossRef]

29. Hundsberger, T.; Schoser, B.; Leupold, D.; Rösler, K.M.; Putora, P.M. Comparison of recent pivotal recommendations for the diagnosis and treatment of late-onset Pompe disease using diagnostic nodes—the Pompe disease burden scale. *J. Neurol.* **2019**, *266*, 2010–2017. [CrossRef]

30. Kronn, D.F.; Day-Salvatore, D.; Hwu, W.-L.L.; Jones, S.A.; Nakamura, K.; Okuyama, T.; Swoboda, K.J.; Kishnani, P.S.; Pompe Disease Newborn Screening Working Group. Management of confirmed newborn-screened patients with pompe disease across the disease spectrum. *Pediatrics* **2017**, *140*, S24–S45. [CrossRef]

31. Lagler, F.B.; Moder, A.; Rohrbach, M.; Hennermann, J.; Mengel, E.; Gökce, S.; Hundsberger, T.; Rösler, K.M.; Karabul, N.; Huemer, M. Extent, impact, and predictors of diagnostic delay in Pompe disease: A combined survey approach to unveil the diagnostic odyssey. *JIMD Rep.* **2019**, *49*, 89–95. [CrossRef]

32. Chien, Y.H.; Lee, N.C.; Huang, H.J.; Thurberg, B.L.; Tsai, F.J.; Hwu, W.L. Later-onset pompe disease: Early detection and early treatment initiation enabled by newborn screening. *J. Pediatr.* **2011**, *158*, 1023–1027.e1. [CrossRef]

33. De Groot, A.S.; Kazi, Z.B.; Martin, R.F.; Terry, F.E.; Desai, A.K.; Martin, W.D.; Kishnani, P.S. HLA- and genotype-based risk assessment model to identify infantile onset pompe disease patients at high-risk of developing significant anti-drug antibodies (ADA). *Clin. Immunol.* **2019**, *200*, 66–70. [CrossRef] [PubMed]

34. Desai, A.K.; Li, C.; Rosenberg, A.S.; Kishnani, P.S. Immunological challenges and approaches to immunomodulation in Pompe disease: A literature review. *Ann. Transl. Med.* **2019**, *7*, 285. [CrossRef]

35. Labrousse, P.; Chien, Y.-H.H.; Pomponio, R.J.; Keutzer, J.; Lee, N.-C.C.; Akmaev, V.R.; Scholl, T.; Hwu, W.-L.L. Genetic heterozygosity and pseudodeficiency in the Pompe disease newborn screening pilot program. *Mol. Genet. Metab.* **2010**, *99*, 379–383. [CrossRef] [PubMed]

36. Chiang, S.-C.C.; Hwu, W.-L.L.; Lee, N.-C.C.; Hsu, L.-W.W.; Chien, Y.-H.H. Algorithm for Pompe disease newborn screening: Results from the Taiwan screening program. *Mol. Genet. Metab.* **2012**, *106*, 281–286. [CrossRef]

37. Liao, H.-C.; Chiang, C.-C.; Niu, D.-M.; Wang, C.-H.; Kao, S.-M.; Tsai, F.-J.; Huang, Y.-H.; Liu, H.-C.; Huang, C.-K.; Gao, H.-J.; et al. Detecting multiple lysosomal storage diseases by tandem mass spectrometry—A national newborn screening program in Taiwan. *Clin. Chim. Acta* **2014**, *431*, 80–86. [CrossRef] [PubMed]

38. Chiang, S.-C.; Chen, P.-W.; Hwu, W.-L.; Lee, A.-J.; Chen, L.-C.; Lee, N.-C.; Chiou, L.-Y.; Chien, Y.-H. Performance of the four-plex tandem mass spectrometry lysosomal storage disease newborn screening test: The necessity of adding a 2nd tier test for Pompe disease. *Int. J. Neonatal Screen.* **2018**, *4*, 41. [CrossRef]

39. Scott, C.R.; Elliott, S.; Buroker, N.; Thomas, L.I.; Keutzer, J.; Glass, M.; Gelb, M.H.; Turecek, F. Identification of infants at risk for developing Fabry, Pompe, or mucopolysaccharidosis-I from newborn blood spots by tandem mass spectrometry. *J. Pediatr.* **2013**, *163*, 498–503. [CrossRef]

40. Elliott, S.; Buroker, N.; Cournoyer, J.J.; Potier, A.M.; Trometer, J.D.; Elbin, C.; Schermer, M.J.; Kantola, J.; Boyce, A.; Turecek, F.; et al. Pilot study of newborn screening for six lysosomal storage diseases using Tandem Mass Spectrometry. *Mol. Genet. Metab.* **2016**, *118*, 304–309. [CrossRef] [PubMed]

41. Hopkins, P.V.; Campbell, C.; Klug, T.; Rogers, S.; Raburn-Miller, J.; Kiesling, J. Lysosomal Storage Disorder Screening Implementation: Findings from the First Six Months of Full Population Pilot Testing in Missouri. *J. Pediatr.* **2015**, *166*, 172–177. [CrossRef] [PubMed]

42. Burton, B.K.; Charrow, J.; Hoganson, G.E.; Waggoner, D.; Tinkle, B.; Braddock, S.R.; Schneider, M.; Grange, D.K.; Nash, C.; Shryock, H.; et al. Newborn Screening for Lysosomal Storage Disorders in Illinois: The Initial 15-Month Experience. *J. Pediatr.* **2017**, *190*, 130–135. [CrossRef] [PubMed]

43. Wasserstein, M.P.; Caggana, M.; Bailey, S.M.; Desnick, R.J.; Edelmann, L.; Estrella, L.; Holzman, I.; Kelly, N.R.; Kornreich, R.; Kupchik, S.G.; et al. The New York pilot newborn screening program for lysosomal storage diseases: Report of the First 65,000 Infants. *Genet. Med.* **2019**, *21*, 631–640. [CrossRef] [PubMed]

44. Mechtler, T.P.; Stary, S.; Metz, T.F.; De Jesús, V.R.; Greber-Platzer, S.; Pollak, A.; Herkner, K.R.; Streubel, B.; Kasper, D.C. Neonatal screening for lysosomal storage disorders: Feasibility and incidence from a nationwide study in Austria. *Lancet* **2012**, *379*, 335–341. [CrossRef]

45. Wittmann, J.; Karg, E.; Turi, S.; Legnini, E.; Wittmann, G.; Giese, A.-K.; Lukas, J.; Gölnitz, U.; Klingenhäger, M.; Bodamer, O.; et al. Newborn screening for lysosomal storage disorders in hungary. *JIMD Rep.* **2012**, *6*, 117–125.

46. Burlina, A.B.; Polo, G.; Salviati, L.; Duro, G.; Zizzo, C.; Dardis, A.; Bembi, B.; Cazzorla, C.; Rubert, L.; Zordan, R.; et al. Newborn screening for lysosomal storage disorders by tandem mass spectrometry in North East Italy. *J. Inherit. Metab. Dis.* **2018**, *41*, 209–219. [CrossRef]

47. Navarrete-Martínez, J.I.; Limón-Rojas, A.E.; de Jesús Gaytán-García, M.; Reyna-Figueroa, J.; Wakida-Kusunoki, G.; del Rocío Delgado-Calvillo, M.; Cantú-Reyna, C.; Cruz-Camino, H.; Cervantes-Barragán, D.E. Newborn screening for six lysosomal storage disorders in a cohort of Mexican patients: Three-year findings from a screening program in a closed Mexican health system. *Mol. Genet. Metab.* **2017**, *121*, 16–21.

48. Bravo, H.; Neto, E.C.; Schulte, J.; Pereira, J.; Filho, C.S.; Bittencourt, F.; Sebastião, F.; Bender, F.; de Magalhães, A.P.S.; Guidobono, R.; et al. Investigation of newborns with abnormal results in a newborn screening program for four lysosomal storage diseases in Brazil. *Mol. Genet. Metab. Rep.* **2017**, *12*, 92–97. [CrossRef]

49. Chien, Y.-H.; Chiang, S.-C.; Zhang, X.K.; Keutzer, J.; Lee, N.-C.; Huang, A.-C.; Chen, C.-A.; Wu, M.-H.; Huang, P.-H.; Tsai, F.-J.; et al. Early Detection of Pompe Disease by Newborn Screening Is Feasible: Results from the Taiwan Screening Program. *Pediatrics* **2008**, *122*, e39–e45. [CrossRef]

50. Nakamura, K.; Hattori, K.; Endo, F. Newborn screening for lysosomal storage disorders. *Am. J. Med. Genet. Part C Semin. Med. Genet.* **2011**, *157*, 63–71. [CrossRef]

51. Sawada, T.; Kido, J.; Yoshida, S.; Sugawara, K.; Momosaki, K.; Inoue, T.; Tajima, G.; Sawada, H.; Mastumoto, S.; Endo, F.; et al. Newborn screening for Fabry disease in the western region of Japan. *Mol. Genet. Metab. Rep.* **2020**, *22*, 100562. [CrossRef] [PubMed]

52. Prosser, L.A.; Lam, K.K.; Grosse, S.D.; Casale, M.; Kemper, A.R. Using Decision Analysis to Support Newborn Screening Policy Decisions: A Case Study for Pompe Disease. *MDM Policy Pract.* **2018**, *3*. [CrossRef] [PubMed]

53. Paciotti, S.; Persichetti, E.; Pagliardini, S.; Deganuto, M.; Rosano, C.; Balducci, C.; Codini, M.; Filocamo, M.; Menghini, A.R.; Pagliardini, V.; et al. First pilot newborn screening for four lysosomal storage diseases in an Italian region: Identification and analysis of a putative causative mutation in the *GBA* gene. *Clin. Chim. Acta* **2012**, *413*, 1827–1831. [CrossRef] [PubMed]

54. Camargo Neto, E.; Schulte, J.; Pereira, J.; Bravo, H.; Sampaio-Filho, C.; Giugliani, R. Neonatal screening for four lysosomal storage diseases with a digital microfluidics platform: Initial results in Brazil. *Genet. Mol. Biol.* **2018**, *41*, 414–416. [CrossRef]

55. Rojas Málaga, D.; Brusius-Facchin, A.C.; Michelin-Tirelli, K.; Félix, T.M.; Schulte, J.; Pereira, J.; Camargo Neto, E.; Sampaio Filho, C.; Giugliani, R. Pseudo deficiency of acid α-glucosidase: A challenge in the newborn screening for Pompe disease. *Genet. Mol. Res.* **2017**, *16*, 16039844.

56. Ausems, M.G.E.M.; Verbiest, J.; Hermans, M.M.P.; Kroos, M.A.; Beemer, F.A.; Wokke, J.H.J.; Sandkuijl, L.A.; Reuser, A.J.J.; Van Der Ploeg, A.T. Frequency of glycogen storage disease type II in The Netherlands: Implications for diagnosis and genetic counselling. *Eur. J. Hum. Genet.* **1999**, *7*, 713–716. [CrossRef]

57. Becker, J.A.; Vlach, J.; Raben, N.; Nagaraju, K.; Adams, E.M.; Hermans, M.M.; Reuser, A.J.; Brooks, S.S.; Tifft, C.J.; Hirschhorn, R.; et al. The African origin of the common mutation in African American patients with glycogen-storage disease type II. *Am. J. Hum. Genet.* **1998**, *62*, 991–994. [CrossRef]

58. Herbert, M.; Cope, H.; Li, J.S.; Kishnani, P.S. Severe Cardiac Involvement Is Rare in Patients with Late-Onset Pompe Disease and the Common c.-32-13T>G Variant: Implications for Newborn Screening. *J. Pediatr.* **2018**, *198*, 308–312. [CrossRef]

59. Pruniski, B.; Lisi, E.; Ali, N. Newborn screening for Pompe disease: Impact on families. *J. Inherit. Metab. Dis.* **2018**, *41*, 1189–1203. [CrossRef]

60. Prater, S.N.; Banugaria, S.G.; Dearmey, S.M.; Botha, E.G.; Stege, E.M.; Case, L.E.; Jones, H.N.; Phornphutkul, C.; Wang, R.Y.; Young, S.P.; et al. The emerging phenotype of long-term survivors with infantile Pompe disease. *Genet. Med.* **2012**, *14*, 800–810. [CrossRef]

61. Chien, Y.H.; Lee, N.C.; Peng, S.F.; Hwu, W.L. Brain development in infantile-onset pompe disease treated by enzyme replacement therapy. *Pediatr. Res.* **2006**, *60*, 349–352. [CrossRef]

62. Spiridigliozzi, G.A.; Keeling, L.A.; Stefanescu, M.; Li, C.; Austin, S.; Kishnani, P.S. Cognitive and academic outcomes in long-term survivors of infantile-onset Pompe disease: A longitudinal follow-up. *Mol. Genet. Metab.* **2017**, *121*, 127–137. [CrossRef] [PubMed]

63. Roig-Zamboni, V.; Cobucci-Ponzano, B.; Iacono, R.; Ferrara, M.C.; Germany, S.; Bourne, Y.; Parenti, G.; Moracci, M.; Sulzenbacher, G. Structure of human lysosomal acid α-glucosidase—A guide for the treatment of Pompe disease. *Nat. Commun.* **2017**, *8*, 1111. [CrossRef] [PubMed]

64. Lim, J.-A.; Yi, H.; Gao, F.; Raben, N.; Kishnani, P.S.; Sun, B. Intravenous Injection of an AAV-PHP.B Vector Encoding Human Acid α-Glucosidase Rescues Both Muscle and CNS Defects in Murine Pompe Disease. *Mol. Ther. Methods Clin. Dev.* **2019**, *12*, 233–245. [CrossRef] [PubMed]

65. Byrne, B.J.; Fuller, D.D.; Smith, B.K.; Clement, N.; Coleman, K.; Cleaver, B.; Vaught, L.; Falk, D.J.; McCall, A.; Corti, M. Pompe disease gene therapy: Neural manifestations require consideration of CNS directed therapy. *Ann. Transl. Med.* **2019**, *7*, 290. [CrossRef] [PubMed]

International Journal of
Neonatal Screening

Article

Newborn Screening for Pompe Disease in Illinois: Experience with 684,290 Infants

Barbara K. Burton [1,2,*], Joel Charrow [1,2], George E. Hoganson [3], Julie Fleischer [4], Dorothy K. Grange [5], Stephen R. Braddock [6], Lauren Hitchins [2], Rachel Hickey [2], Katherine M. Christensen [6], Daniel Groepper [4], Heather Shryock [7], Pamela Smith [7], Rong Shao [8] and Khaja Basheeruddin [8]

[1] Department of Pediatrics, Feinberg School of Medicine of Northwestern University, Chicago, IL 60611, USA; JCharrow@luriechildrens.org
[2] Department of Pediatrics, Ann and Robert H. Lurie Children's Hospital of Chicago, Chicago, IL 60611, USA; lhitchins@luriechildrens.org (L.H.); rahickey@luriechildrens.org (R.H.)
[3] Department of Pediatrics, University of Illinois College of Medicine, Chicago, IL 60612, USA; geh@uic.edu
[4] Department of Pediatrics, Southern Illinois University School of Medicine, Springfield, IL 62701, USA; jfleischer95@siumed.edu (J.F.); dgroepper@siumed.edu (D.G.)
[5] Department of Pediatrics, Washington University School of Medicine and St. Louis Children's Hospital, St. Louis, MO 63110, USA; grange_d@kids.wustl.edu
[6] Department of Pediatrics, Saint Louis University, St. Louis, MO 63104, USA; braddock@slu.edu (S.R.B.); katherine.christensen@health.slu.edu (K.M.C.)
[7] Office of Health Promotion, Illinois Department of Public Health, Springfield, IL 62761, USA; Heather.Shryock@Illinois.gov (H.S.); Pamela.E.Smith@illinois.gov (P.S.)
[8] Newborn Screening Laboratory, Illinois Department of Public Health, Chicago, IL 60603, USA; Rong.Shao@Illinois.gov (R.S.); Khaja.Basheeruddin@Illinois.gov (K.B.)
* Correspondence: bburton@luriechildrens.org

Received: 11 December 2019; Accepted: 18 January 2020; Published: 21 January 2020

Abstract: Statewide newborn screening for Pompe disease began in Illinois in 2015. As of 30 September 2019, a total of 684,290 infants had been screened and 395 infants (0.06%) were screen positive. A total of 29 cases of Pompe disease were identified (3 infantile, 26 late-onset). While many of the remainder were found to have normal alpha-glucosidase activity on the follow-up testing (234 of 395), other findings included 62 carriers, 39 infants with pseudodeficiency, and eight infants who could not be given a definitive diagnosis due to inconclusive follow-up testing.

Keywords: Pompe disease; newborn screening

1. Introduction

Pompe disease is an autosomal recessive lysosomal storage disorder resulting from the deficiency of acid alpha-glucosidase (GAA). The deficiency of enzyme activity results in the lysosomal accumulation of glycogen and multisystemic clinical manifestations, including prominent skeletal muscle weakness. Patients with the most severe form of the disorder, referred to as infantile onset Pompe disease (IOPD), also have cardiac involvement manifested as hypertrophic cardiomyopathy. In the absence of treatment, patients with IOPD rarely survive beyond two years of age. Patients with late onset Pompe disease (LOPD) may develop clinical manifestations at any age from early childhood through adult life. Progressive limb girdle muscle weakness is the hallmark of the disorder, with disproportionate involvement of the respiratory muscles often leading to respiratory insufficiency. Cardiac involvement is rare in patients with LOPD. There are patients who exhibit an intermediate phenotype with onset of muscle weakness and motor delay in the first year of life without cardiomyopathy. These patients are variably referred to as having either LOPD or atypical IOPD. Enzyme replacement therapy with alglucosidase alfa has been available for both IOPD and LOPD since 2006 and has significantly changed the natural history of the disorder. In patients with IOPD, it prolongs ventilator-free survival and often results in resolution of cardiomyopathy [1]. A subset of patients with IOPD, particularly those who are cross-reacting material (CRIM) negative, are at high risk of developing high titer antibodies to the enzyme, however, with subsequent loss of efficacy [2]. Immune modulation protocols have been developed and have shown to be effective in preventing the development of high titer antibodies in these patients [3]. In patients with LOPD, treatment with alglucosidase alfa may result in improved endurance as measured on the 6-min walk test and stabilization of motor and pulmonary function [4].

Patients with IOPD invariably exhibit elevated levels of serum creatine kinase (CK), whether diagnosed clinically or through newborn screening [5]. Patients with LOPD often, but not always, have elevated CK levels as well. In all patients with elevated levels, a decline is typically observed after initiation of enzyme replacement therapy. A second biomarker that is useful in the diagnosis and monitoring of Pompe disease is a specific urinary glucose tetrasaccharide (Glc4 or Hex4) [6]. In the newborn screening setting, levels of this tetrasaccharide have been reported to be consistently elevated in patients with IOPD but normal in those with LOPD who do not require therapy prior to three years of age [7].

Pseudodeficiency for the GAA enzyme has been well-described and is particularly common in Asian populations [8]. Patients with pseudodeficiency have low levels of GAA measured in vitro, at times as low as those observed in affected individuals but have no evidence of clinical disease or glycogen storage in tissues. Common alleles associated with pseudodeficiency have been identified.

The rapidly progressive nature of IOPD and the availability of disease-modifying therapy was the impetus for the initiation of a pilot screening program for Pompe disease in Taiwan in 2005. Newborn screening for this disorder was shown to not only be feasible [9], but to also improve the prognosis for infantile onset Pompe disease through the earlier implementation of enzyme replacement therapy [10]. Missouri became the first state in the United States to implement newborn screening for Pompe disease in 2013 using the digital microfluidic method [11]. The data from both Taiwan and Missouri were instrumental in securing the addition of Pompe disease to the Recommended Uniform Screening Panel (RUSP) in the US in 2015.

Illinois was the second state in the United Sates to implement statewide newborn screening for Pompe disease and the first to do so using tandem mass spectrometry. Pilot screening in selected hospitals began in November 2014 and was expanded statewide in June 2015. Results from the initial 15 months of screening were previously reported [12]. The purpose of this communication was to extend the initial report and describe the outcome of Pompe newborn screening through September 2019.

2. Materials and Methods

Newborn screening for Pompe disease is performed by determination of alpha-glucosidase enzyme activity in dried blood spots by liquid chromatography–mass spectrometry (LC–MS) using a multiplex assay with reagents from PerkinElmer®. The method and testing cutoffs have been previously described [12]. All infants born in the state of Illinois have been tested since June 2015. Parents can opt out of any newborn screening but cannot selectively opt out of testing for specific disorders. In practice, the opt-out option is virtually never utilized. Infants with a positive screen are urgently referred to one of several referral centers for diagnostic evaluation. A protocol for follow-up (Figure 1) was developed by a working group prior to the initiation of screening and is generally followed by all of the referral centers but is not mandatory and the specific testing ordered is at the discretion of each consulting provider. All centers obtain blood for creatinine kinase (CK) and alpha-glucosidase activity at the time of initial assessment, as well as cardiac studies including chest radiograph and electrocardiogram with echocardiogram in many cases. When deficient enzyme activity is confirmed, molecular analysis is performed at one of several reference laboratories. When two or more mutations or variants are detected, it is recommended that parental testing be performed for phasing. However, the results of parental testing are not routinely collected by the screening program. Many clinicians obtain urine glucose or hexose tetrasaccharide (Glc4, Hex4) at the initial assessment and some include a full metabolic panel and brain natriuretic peptide (BNP). Over time, most centers have obtained fewer follow-up cardiac assessments than recommended on the algorithm on those infants with no initial evidence of cardiac disease. In particular, most infants who have the common c.-32-13T>G mutation do not undergo additional cardiac evaluation since the incidence of cardiac disease in association with this mutation has been shown to be very low [13]. Follow-up echocardiography for infants with other mutations is variable and at the discretion of the clinician.

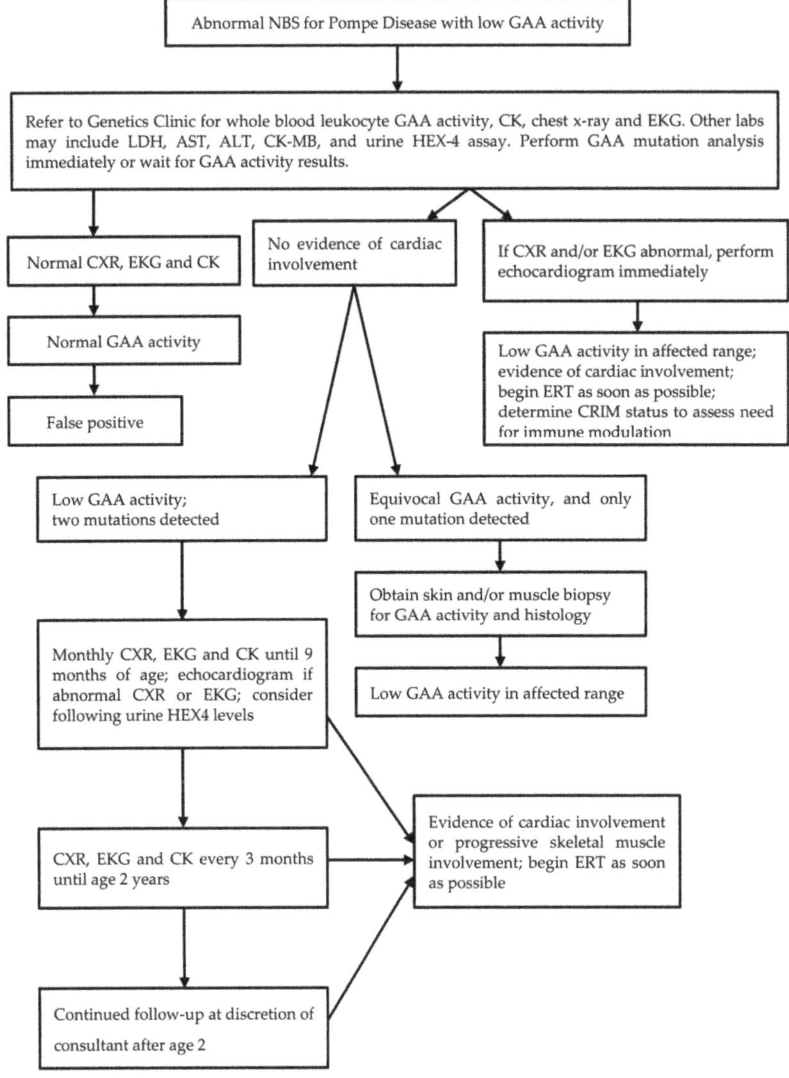

Figure 1. Pompe Disease Algorithm.

Affected infants are classified as having infantile onset Pompe disease (IOPD) if they have evidence of cardiomyopathy in the first year of life. All other definitely affected infants are classified as having late onset Pompe disease (LOPD). The latter category includes infants found to have hypotonia, motor delay or muscle weakness in the first year of life. Infants who have no evidence of cardiac disease but have alpha-glucosidase activity in the affected range and have one pathogenic mutation and one variant of unknown significance (VUS) or two VUS are referred to as having possible LOPD or undetermined status. It is recognized that these infants may not be affected since the variants of unknown significance could be benign, but follow-up is recommended nonetheless, similar to what is provided to infants with definite LOPD.

In the case of infants diagnosed with late onset Pompe disease (LOPD), or possible LOPD defined as low enzyme activity with one pathogenic mutation and one variant of undetermined significance (VUS) or two VUSs in trans, follow-up assessments are typically performed in the first year of life at 3 month intervals and include history and physical examination, developmental assessment, serum CK, and, in some cases, a complete metabolic panel and/or BNP, and urine Glc4. The undetermined cases are followed in the same manner as those with definite LOPD. For patients doing well, the interval between assessments is typically increased to 6 months to a year after one year of age. When LOPD or possible LOPD is diagnosed in an infant, older siblings born prior to the onset of newborn screening are routinely tested.

3. Results

A total of 684,290 infants were screened between 3 November 2014 and 30 September 2019. Three hundred and ninety-seven of these (0.06%), or 1 in 1724, screened positive for Pompe disease and were referred for diagnostic assessment. The outcome of the evaluation of these infants is seen in Table 1. A total of 29 infants had a definitive diagnosis of Pompe disease for a minimum incidence in the population of 1 in 23,596. Of these 29, 26 infants (90%) were found to have late onset Pompe disease. An additional 8 infants were defined as having an undetermined phenotype or "possible" Pompe disease. These were infants with low enzyme activity and two variants in the GAA gene, one or both of which were of unknown significance. Since it is likely that some of these infants will go on to develop Pompe disease, the incidence in the population is likely higher than the 1 in 23,596 calculated from the definite cases. Since screening was initiated, there have been no reports of infants diagnosed with Pompe disease who were missed by newborn screening. There have also been no infants with GAA activity in the affected range with only a single variant detected in the gene.

Table 1. Outcome of follow-up in infants screen positive for Pompe disease (*n* = 395).

Category	Number of Infants Identified
Infantile Pompe disease	3
Late onset Pompe disease	26
Normal enzyme activity	234
Carrier [1]	62
Pseudodeficiency [2]	39
Phenotype undetermined [3]	8
Loss to follow-up or refused	7
Died prior to follow-up [4]	1
Pending	15

[1] Infants with one pathogenic variant, or one VUS, with or without pseudodeficiency alleles were classified as carriers. [2] Infants in this category had only pseudodeficiency allelesc.1726G>A, c.2065G>A, and/or c.271G>A. 38 of 39 infants in this group were of Asian descent. [3] Infants in this category had one pathogenic variant and one or two VUS. [4] This was a premature infant who had multiple complications of prematurity but no findings to suggest Pompe disease.

Three infants had evidence of cardiomyopathy at the time of their initial assessment and were diagnosed with infantile onset Pompe disease (IOPD). All three were determined to be CRIM positive based on molecular analysis. Two were started on enzyme replacement therapy (ERT) with no immune modulation at 4 and 6 weeks of age. The third infant was also CRIM positive but had a mutation previously found to confer high risk for the development of high titer antibodies and had very severe cardiomyopathy requiring treatment in an intensive care unit from day 1 of life. He was treated with ERT beginning at 10 days of age and was simultaneously treated with rituximab, methotrexate, and intravenous immunoglobulin [14]. All three infants are currently doing well, all are ambulatory, and none of the three required any ventilatory support at the last follow-up. There were no infants who had a normal cardiac assessment initially who were later found to have evidence of cardiomyopathy.

Thus far, only one infant with LOPD has been started on ERT (Table 2, Case 25). This was an infant who, at the time of initial assessment, was found to have a CK of 555 international units per liter (IU/L) but no evidence of cardiomyopathy. At two months of age, head lag and mild hypotonia were noted but physical therapy assessment revealed a normal Alberta infant motor scale (AIMS) score at the 85th percentile for age. Multiple additional measurements of CK continued to be in the range of 550–650 U/L. Development appeared to progress normally for several months but by 9 months of age, significant hip laxity and gross motor delay were evident. The AIMS score had decreased to below the 5th percentile for age. A decision was made to initiate treatment and ERT was started at ten months of age, along with short term methotrexate [15]. After three months of therapy, the patient was making motor progress but continued to exhibit significant weakness so the dose of alglucosidase alfa was increased from the label dose of 20 mg/kg every two weeks to 40 mg/kg every two weeks. No other infants or children identified through newborn screening as having LOPD have been reported to have persistent clinical symptoms, although elevations of CK are common at diagnosis, as seen in Table 2.

Table 2. Follow-up data on infants with definite or "possible" Pompe disease.

Case	Genotype	Phenotype	GAA Activity [a] Result (nl)	CK (IU/L) Result (nl)	Urine Glc4 or Hex4 [b] Result (nl)	Cardiac Findings	Other Clinical Findings
1	c.2560C>T, c.1211A>T, c.2161G>C	IOPD	0.02 (>3.0)	1064 (35–232)	NR	HCM	Hypotonia
2	c.1437+1G>A, c.2227C>T	IOPD	0.8 (>3.88)	566 (32–250)	NR	HCM	Hypotonia; Motor delay
3	c.2560C>T, c.2459_2461del	IOPD	1.6 (>3.88)	3488 (30–279)	Glc4 14.9 (0.14–1.29)	HCM	Initial hypotonia [c]
4–16	c.-32-13T>G homozygous	LOPD	0.0–2.8	153–669 (8/17 elevated)	See footnote [d]	Normal [e]	None
17	c.-32-13T>G, c.1655T>C	LOPD	0.8 (>3.0)	550	Normal	ASD	Mild hypotonia
18	c.-32-13T>G, c.2238G>C	LOPD	2.55 (>3.88)	86 (29–168)	Normal	None	None
19	c.-32-13T>G, c.1839G>A	LOPD	1.5 (>3.88)	641 (30–279)	Glc4 7.59 (0.14–1.29)	None	None
20	c.-32-13T>G, c.258DPC	LOPD	0.3 (>3.0)	NR	NR	None	None
21	c.-32-13T>G, c.2238G>C, c.2065G>A	LOPD	1.0 (>3.0)	NR	Hex4 41.6 (<20)	RVH on ECG; PFO on echo	None
22	c.-32-13T>G, c.2297A>G	LOPD [f]	2.3 (>3.88)	168 (55–170)	Glc4 1.21 (0.08–1.37)	Normal	None
23	c.307T>G, c.1375G>C, c.271G>A	LOPD [f]	1.6 (>3.88)	Normal	Glc4 2.0 (1.14–1.29)	Normal	None
24	c.1637-3_1637-4delinsG, c.1831G>A	LOPD	2.4 (>3.88)	93 (30–279)	Glc4 12.98 (0.14–1.29)	Normal	None

Table 2. *Cont.*

Case	Genotype	Phenotype	GAA Activity [a] Result (nl)	CK (IU/L) Result (nl)	Urine Glc4 or Hex4 [b] Result (nl)	Cardiac Findings	Other Clinical Findings
25	c.-32-12T>G, c.2219-2220delTG	LOPD [g]	2.0 (>3.88)	555 (30–279)	Glc4 11.79 (0.14–1.29)	PFO	Hypotonia; gross motor delay
26	c.2238G>C, c.2242dupG	LOPD [f]	2.9 (>3.88)	142 (30–279)	Glc4 6.54 (0.14–1.29)	Normal	None
27	c.2173delC, c.858+17-858+23delCGGGGCGG	LOPD	2.9 (>3.88)	272 (30–279)	NR	Normal	None
28	c.1121G>T, c.885C>T	LOPD	0.3 (>3.0)	NR	NR	Normal	None
29	c.307T>G, c.525delT	LOPD	0.5 (>3.0)	193 (55–170)	NR	Normal	None
30	c.655G>A, c.1418G>C	UND	3.0 (>7.4)	73 (39–308)	Hex4 11.2 (<20)	PFO	None
31	c.525delT, c.265C>T	UND [f]	2.9 (>3.88)	74 (30–279)	Glc4 6.76 (0.14–1.29)	PFO [h]	None
32	c.1942G>A, c.1346C>T, c.2065G>A, c.1726G>A	UND	0.2 (>3.0)	167 (39–308)	NR	Normal	None
33	c.664G>A, c.1346C>T	UND	0.0 (>3.88)	101 (30–279)	NR	Normal	None
34	c.726G>A, c.1357G>A	UND	0.7 (>3.0)	NR	NR	Normal	None
35	c.1631T>A, c.2509C>T, c.2065G>A	UND	UND	0.7 (>3.0)	NR	NR	Normal
36	c.307T>G, c.265C>T	UND	2.1 (>3.88)	152 (30–279)	NR	NR	Normal
37	c.1781G>A, c.1194+3G>C	UND	3.5 (>3.88)	571 (30–279)	Glc4 3.46 (0.14–1.29)	Normal	None

[a] GAA activity units: pmol/punch/h. [b] Urine Glc4 or Hex4 activity units: mmol/mol creatinine. [c] Initial hypotonia and hypoventilation during sleep; resolved by 10 months of age. [d] 13 Cases: NR for 7; Glc4 5.42–9.83 (0.142–1.29) for 4; Hex4 6.8 and 13.2 (<20) for 2. [e] One case with left ventricular hypertrophy on ECG, with normal echo; all others normal. [f] Older sibling, born prior to newborn screening, has same genotype. [g] Symptomatic at 9 months of age, started on ERT. [h] Also had mildly dilated ascending aorta. ASD = Atrial septal defect. CK = creatine kinase. GAA activity = dried blood spot alpha-glucosidase activity determined at time of diagnostic evaluation. Glc4 = glucose tetrasaccharide. HCM = Hypertrophic cardiomyopathy. Hex4 = hexose tetrasaccharide. IOPD = Infantile onset Pompe disease. LOPD = Late onset Pompe disease. nl = Normal. NR = Not reported. PFO = Patent foramen ovale. RVH = Right ventricular hypertrophy. UND = Undetermined.

For patients with more than two variants, clinical significance of all variants was not always evident. There were patients who had two pathogenic variants and either pseudodeficiency alleles or variants of unknown significance. All of the variants in the patients with definite or possible Pompe disease are listed in Table 2.

The results of the clinical and laboratory assessments performed on affected infants at the time of initial evaluation are seen in Table 2. It should be noted that these data are obtained from the consulting physicians by the Illinois department of public health and reporting may be incomplete in some cases. Due to privacy restrictions restricting release of identifying information, original records on all of these infants were not available to the authors. If a result is listed as NR (not reported), it may not have been done or may simply not have been reported, even if it was done. After the initial data were gathered following diagnosis, a follow-up questionnaire was sent annually to consulting physicians for each patient. The length of follow-up for the patients reported ranges from several months to 4 years.

4. Discussion

Pompe disease newborn screening has been successfully implemented in the state of Illinois over the past 5 years. As reported from other states [16], the overall incidence of Pompe disease was found to be higher than previously estimated, with the large majority of cases being of the late onset type. In addition, there were patients in whom a definitive diagnosis could not be established, who will require ongoing follow-up.

The patients identified through newborn screening who are now on treatment have thus far had an excellent outcome and experience elsewhere has documented the benefits of treatment initiation prior to the onset of overt clinical symptoms [17]. Questions remain regarding the optimal follow-up of infants diagnosed with LOPD and the optimal time for initiation of treatment. It is anticipated that the collective experience of the many states now performing screening will shed light on these issues in the coming years. In the meantime, a method of systemic data collection to enable long-term assessment of outcomes and best practices is critically needed. Although there is an industry-sponsored registry for patients with confirmed Pompe disease [18], it is not ideal for the inclusion of asymptomatic patients, particularly those with genotypes that do not permit a definitive diagnosis ("possible" Pompe patients) and is not accessible by all treating physicians. A similar need exists for other disorders, including other lysosomal disorders, recently added to newborn screening since many are associated with phenotypes that may not become evident until much later in childhood or even in adult life.

5. Conclusions

Pompe newborn screening in the state of Illinois has been ongoing for 5 years and has led to the diagnosis of 3 infants with IOPD, 26 with LOPD, and eight with an undetermined diagnosis.

Author Contributions: B.K.B., J.C., G.E.H., J.F., S.R.B., D.K.G., R.H., L.H., D.G., and K.M.C. were involved in evaluation of patients with positive newborn screening tests and in treatment of those patients with a definitive diagnosis. P.S. and H.S. obtained and collated follow-up data on infants with a positive newborn screening test. R.S. and K.B. performed and analyzed the newborn screening tests. B.K.B. curated data obtained from the Department of Public Health and wrote the first draft of the manuscript. All authors have read and agreed to the published version of the manuscript.

Funding: This research received no external funding.

Acknowledgments: The authors would like to acknowledge Vera Shively for her expert editorial assistance.

Conflicts of Interest: B.K.B. has received consulting fees and/or honoraria from Sanofi Genzyme, Biomarin, Shire (a Takeda company), Ultragenyx, Moderna, Horizon, Alexion, Chiesi, Denali, JCR Pharma, Aeglea, and Agios. She has conducted contracted research for Sanofi Genzyme, Biomarin, Shire, Alexion, Ultragenyx, and Sangamo and Homology Medicines. J.C. has also received grant support for research from: Sanofi Genzyme, Shire, Amicus, BioMarin, and Protalix. Dorothy K. Grange conducts contracted research for Biomarin and Shire (a Takeda company). The following authors declare no conflict of interest: K.B., S.R.B., K.M.C., J.F., D.G., R.H., L.H., G.E.H., H.S., R.S., and P.S.

References

1. Kishnani, P.S.; Corzo, D.; Leslie, N.D.; Gruskin, D.; van der Ploeg, A.; Clancy, J.P.; Parini, R.; Morin, G.; Beck, M.; Bauer, M.S.; et al. Early treatment with alglucosidase alfa prolongs long-term survival in infants with Pompe disease. *Pediatr. Res.* **2009**, *66*, 329–335. [CrossRef] [PubMed]
2. Kishnani, P.S.; Goldenberg, P.C.; DeArmey, S.L.; Heller, J.; Benjamin, D.; Young, S.; Bali, D.; Smith, S.A.; Li, J.S.; Mandel, H.; et al. Cross-reactive immunologic material status affects treatment outcomes in Pompe disease infants. *Mol. Genet. Metab.* **2010**, *99*, 26–33. [CrossRef] [PubMed]
3. Messinger, Y.H.; Mendelsohn, N.J.; Rhead, W.; Dimmock, D.; Hershkovitz, E.; Champion, M.; Jones, S.A.; Olson, R.; White, A.; Wells, C.; et al. Successful immune tolerance induction to enzyme replacement therapy in CRIM-negative infants with Pompe disease. *Mol. Genet. Metab.* **2012**, *14*, 135–142.
4. Van der Ploeg, A.T.; Clemens, P.R.; Corzo, D.; Escolar, D.M.; Florence, J.; Groeneveld, G.J.; Herson, S.; Kishnani, P.S.; Laforet, P.; Lake, S.L.; et al. A randomized study of alglucosidase alfa in late-onset Pompe disease. *N. Engl. J. Med.* **2010**, *362*, 1396–1406. [CrossRef] [PubMed]

5. Kishnani, P.S.; Steiner, R.D.; Bali, D.; Berger, K.; Byrne, B.J.; Case, L.E.; Crowley, J.F.; Downs, S.; Howell, R.R.; Kravitz, R.M.; et al. Pompe disease diagnosis and management guidelines. *Genet. Med.* **2006**, *8*, 267–288. [CrossRef] [PubMed]

6. Young, S.P.; Piraud, M.; Goldstein, J.L.; Zhang, H.; Rehder, C.; Laforet, P.; Kishnani, P.S.; Millington, D.S.; Bashir, M.R.; Bali, D.S. Assessing disease severity in Pompe disease: The roles of a urine glucose tetrasaccharide biomarker and imaging techniques. *Am. J. Med. Genet. C Semin. Med. Genet.* **2012**, *160C*, 50–58. [CrossRef] [PubMed]

7. Chien, Y.H.; Goldstein, J.L.; Hwu, W.L.; Smith, P.B.; Lee, N.C.; Chiang, S.C.; Tolun, A.A.; Zhang, H.; Vaisnins, A.E.; Millington, D.S.; et al. Baseline urine glucose tetrasaccharide concentrations in patients with infantile- and late-onset Pompe disease identified by newborn screening. *JIMD Rep.* **2015**, *19*, 67–73. [PubMed]

8. Kumamoto, S.; Katafuchi, T.; Nakamura, K.; Endo, F.; Oda, E.; Okuyama, T.; Kroos, M.A.; Reuser, A.J.; Okumiya, T. High frequency of acid alpha-glucosidase pseudodeficiency complicating newborn screening for glycogen storage disease type II in the Japanese population. *Mol. Genet. Metab.* **2009**, *99*, 190–195. [CrossRef] [PubMed]

9. Chien, Y.H.; Chiang, S.C.; Zhang, X.K.; Keutzer, J.; Lee, N.C.; Huang, A.C.; Chen, C.A.; Wu, M.H.; Huang, P.H.; Tsai, F.J.; et al. Early detection of Pompe disease by newborn screening is feasible: Results from the Taiwan screening program. *Pediatrics* **2008**, *122*, e39–e45. [CrossRef] [PubMed]

10. Chien, Y.H.; Lee, N.C.; Thurberg, B.L.; Chiang, S.C.; Zhang, X.K.; Keutzer, J.; Huang, A.C.; Wu, M.H.; Huang, P.H.; Tsai, F.J.; et al. Pompe disease in infants: Improving the prognosis by newborn screening and early treatment. *Pediatrics* **2009**, *124*, e1116–e1125. [CrossRef] [PubMed]

11. Hopkins, P.V.; Campbell, C.; Klug, T.; Rogers, S.; Raburn-Miller, J.; Kiesling, J. Lysosomal storage disorder newborn screening implementation: Findings from the first six months of full population pilot testing in Missouri. *J. Pediatr.* **2015**, *166*, 172–177. [CrossRef] [PubMed]

12. Burton, B.K.; Charrow, J.; Hoganson, G.E.; Waggoner, D.; Tinkle, B.; Braddock, S.R.; Schneider, M.; Grange, D.K.; Nash, C.; Shryock, H.; et al. Newborn screening for lysosomal storage disorders in Illinois: The initial 15-month experience. *J. Pediatr.* **2017**, *190*, 130–135. [CrossRef] [PubMed]

13. Herbert, M.; Cope, H.; Li, J.S.; Kishnani, P.S. Severe cardiac involvement is rare in patients with late-onset Pompe disease and the common c.-32-13T>G variant: Implications for newborn screening. *J. Pediatr.* **2018**, *198*, 308–312. [CrossRef] [PubMed]

14. Kazi, Z.B.; Desai, A.K.; Berrier, K.L.; Troxler, R.B.; Wang, R.Y.; Abdul-Rahman, O.A.; Tanpaiboon, P.; Mendelsohn, N.J.; Herskovitz, E.; Kronn, D.; et al. Sustained immune tolerance induction in enzyme replacement therapy-treated CRIM-negative patients with infantile Pompe disease. *JCI Insight* **2017**, *2*, e94328. [CrossRef] [PubMed]

15. Kazi, Z.B.; Desai, A.K.; Troxler, R.B.; Kronn, D.; Packman, S.; Sabbadini, M.; Rizzo, W.B.; Scherer, K.; Abdul-Rahman, O.; Tanpaiboon, P.; et al. An immune tolerance approach using transient low-dose methotrexate in ERT-naïve setting of patients treated with a therapeutic protein: Experience in infantile-onset Pompe disease. *Genet. Med.* **2019**, *21*, 887–895. [CrossRef] [PubMed]

16. Hopkins, P.V.; Klug, T.; Vermette, L.; Raburn-Miller, J.; Kiesling, J.; Rogers, S. Incidence of 4 lysosomal storage disorders from 4 years of newborn screening. *JAMA Pediatr.* **2018**, *172*, 696–697. [CrossRef] [PubMed]

17. Yang, C.F.; Yang, C.C.; Liao, H.C.; Huang, L.Y.; Chiang, C.C.; Ho, H.C.; Lai, C.J.; Chu, T.H.; Yang, T.F.; Hsu, T.R.; et al. Very early treatment for infantile-onset Pompe disease contributes to better outcomes. *J. Pediatr.* **2016**, *169*, 174–180. [CrossRef] [PubMed]

18. Reuser, A.J.J.; van der Ploeg, A.T.; Chien, Y.H.; Llerena, J., Jr.; Abbott, M.A.; Clemens, P.R.; Kimonis, V.E.; Leslie, N.; Maruti, S.S.; Sanson, B.J.; et al. GAA variants and phenotypes amount 1,079 patients with Pompe disease; data from the Pompe Registry. *Hum. Mutat.* **2019**, *40*, 2146–2164. [CrossRef] [PubMed]

 International Journal of
Neonatal Screening

The First Year Experience of Newborn Screening for Pompe Disease in California

Hao Tang *, Lisa Feuchtbaum, Stanley Sciortino, Jamie Matteson, Deepika Mathur, Tracey Bishop and Richard S. Olney

Genetic Disease Screening Program, California Department of Public Health, 850 Marina Bay Parkway, MS 8200, USA; lisa.feuchtbaum@cdph.ca.gov (L.F.); stanley.sciortino@cdph.ca.gov (S.S.); jamie.matteson@cdph.ca.gov (J.M.); deepika.mathur@cdph.ca.gov (D.M.); tracey.bishop@cdph.ca.gov (T.B.); richard.olney@cdph.ca.gov (R.S.O.)
* Correspondence: hao.tang@cdph.ca.gov; Tel.: +1-510-412-1488

Received: 24 December 2019; Accepted: 5 February 2020; Published: 7 February 2020

Abstract: The California Department of Public Health started universal newborn screening for Pompe disease in August 2018 with a two-tier process including: (1) acid alpha-glucosidase (GAA) enzyme activity assay followed by, (2) *GAA* gene sequencing analysis. This study examines results from the first year of screening in a large and diverse screening population. With 453,152 screened newborns, the birth prevalence and GAA enzyme activity associated with various types of Pompe disease classifications are described. The frequency of *GAA* gene mutations and allele variants are reported. Of 88 screen positives, 18 newborns were resolved as Pompe disease, including 2 classic infantile-onset and 16 suspected late-onset form. The c.-32-13T>G variant was the most common pathogenic mutation reported. African American and Asian/Pacific Islander newborns had higher allele frequencies for both pathogenic and pseudodeficiency variants. After the first year of Pompe disease screening in California, the disease distribution in the population is now better understood. With the ongoing long-term follow-up system currently in place, our understanding of the complex genotype-phenotype relationships will become more evident in the future, and this should help us better understand the clinical significance of identified cases.

Keywords: Pompe disease; newborn screening; California

---

1. Introduction

Pompe disease is a sometimes-fatal inherited lysosomal storage disorder caused by the abnormal accumulation of glycogen in cells, which can result in progressive dysfunction of the heart and other muscles. Also known as glycogen storage disease type II, Pompe disease is caused by a deficiency of the acid alpha-glucosidase (GAA) enzyme that breaks down a type of complex sugar, lysosomal glycogen. The birth prevalence of Pompe disease has been estimated to be 1 in 40,000 [1,2], or 25 per 1 million births, although studies from Israel, Taiwan and some parts of the United States reported higher prevalence rates [3–5].

The severity, age of onset, and rate of progression of Pompe disease vary among individuals, who have been generally categorized into three types. The classic infantile-onset Pompe disease (IOPD) shows symptoms within a few months of birth, characterized by fatal cardiomyopathy if untreated. The non-classic infantile-onset form begins before age one, typically with no heart complications. The late-onset Pompe disease (LOPD) appears later in childhood, adolescence, or adulthood [6–8]. The variance in phenotypes has been linked to different *GAA* gene variants, which are the cause for GAA enzyme deficiency. Certain pathogenic variants on both *GAA* alleles severely reduce GAA activity and usually lead to IOPD. On the other hand, some variants of the *GAA* gene exhibit low levels of GAA activity, leading to more moderate forms of Pompe disease [1,9]. To date, enzyme replacement therapy

(ERT) has been the only direct medical treatment for all forms of Pompe disease by reducing GAA deficiency. Treatment beginning as soon as the disease is detected, or as early as possible, can generate the most benefit for patients [10–14].

The clinical work-up of Pompe disease usually involves measuring GAA enzyme activity and molecular analysis to confirm the diagnosis [5,15]. Recent studies have shown that a tandem mass spectrometry (MS/MS)-based GAA enzyme activity assay could be a functional laboratory method for Pompe disease detection [16,17]. The option of multiplex testing for Pompe disease, along with other MS/MS disorders using the same dried blood spot (DBS), helped promote Pompe disease as a viable disorder to add to newborn screening panels [18,19]. The first newborn screening program for Pompe disease was implemented in Taiwan as early as 2005 [20]. Since then, several other countries and U.S. states have conducted pilot screening studies with promising results [5,21–23]. Subsequently, an external condition review workgroup commissioned by the Health Resources and Services Administration examined the evidence for including Pompe disease on the federal Recommended Uniform Screening Panel (RUSP) in 2013 [24]. In 2013, the Advisory Committee on Heritable Disorders in Newborns and Children voted to recommend that the United States Secretary of Health and Human Services add the disorder to the RUSP, which occurred in March 2015 [25]. As of November 2019, 22 states are screening for Pompe disease [26].

The addition of Pompe disease to California's Newborn Screening (NBS) panel followed passage of SB 1095 in the California legislature in 2016 that amended the Health and Safety Code [27,28]. This required the Genetic Disease Screening Program (GDSP) of the California Department of Public Health to add Pompe disease in order to be compliant with the RUSP, and this process has been described in more detail by Bronstein et al. [29]. On August 29, 2018, California began universal screening for Pompe disease.

This paper reports the findings from the first year of population-based Pompe disease screening. We describe our screening and follow-up algorithm as well as epidemiological and clinical outcomes of screening, including disease and variant classification and other characteristics of the Pompe cases identified to date.

2. Materials and Methods

In California, Pompe disease screening is a two-tier process as shown in Figure 1. DBSs are analyzed using flow injection analysis-tandem mass spectrometry (FIA-MS/MS) to measure GAA enzyme activity. Specimens whose GAA enzyme activity levels are below 18% of the daily median are separated into two groups. Those with very low levels, below 10% of the daily median, are immediately called out as screen positive and sent for *GAA* gene sequencing and clinical follow-up, while those with intermediate GAA enzyme levels, between 10% and 18% of the daily median await *GAA* gene sequencing results before the final interpretation is made. The specimens with intermediate GAA enzyme levels are only referred for clinical follow-up if at least one pathogenic variant, likely pathogenic variant or variant of uncertain significance (VUS) is found.

Clinical follow-up is conducted in one of fifteen metabolic specialty care centers across California. The specialty care centers provide genetic counseling, confirmatory testing, diagnosis, and long-term clinical care when appropriate.

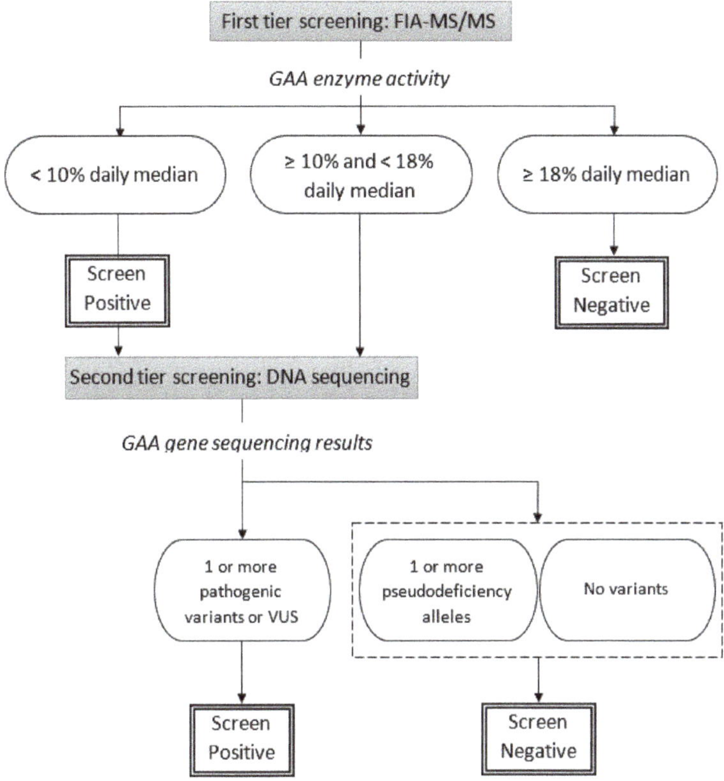

Figure 1. California Pompe disease newborn screening algorithm.

All testing results (biochemical and DNA sequencing), along with demographic information and follow-up reports (short-term follow-up to diagnosis and long-term follow-up for five years), associated with the referred newborns are entered and stored in GDSP's web-based Screening Information System (SIS), including a newborn screening registry that houses all clinically confirmed Pompe disease cases. The categories of the California Pompe disease case resolutions include: (1) classic infantile-onset Pompe disease (with cardiac involvement), (2) non-classic infantile-onset Pompe disease (without cardiac involvement), (3) late-onset Pompe disease, and (4) not-otherwise-specified Pompe disease. After referral, metabolic specialists make the diagnostic decision following established case definitions [30] and general guidelines (Table 1). Newborns who are carriers or who only have pseudodeficiency alleles are also recorded in the registry for reference, but these newborns are not referred for additional clinical follow-up. Variant classification of pathogenic, likely pathogenic, uncertain significance, and pseudodeficiency allele are based on established guidelines with published *GAA* mutations [31–33]. For some of the analyses, we combined late-onset and not-otherwise-specified Pompe disease cases into a "suspected late-onset" category due to the similarities of their diagnostic characteristics (both had no symptoms and had similar GAA levels and variants).

Table 1. California newborn screening Pompe disease diagnosis guideline.

Diagnosis	Mutation Status	Symptoms	Long-Term Follow-Up
Pompe–classic infant onset (with cardiac involvement) *	Pathogenic/likely pathogenic/VUS alleles ** ≥ 2	Yes, with positive cardiac involvement	Yes
Pompe–non-classic infant onset (without cardiac involvement) *	Pathogenic/likely pathogenic/VUS alleles ** ≥ 2	Yes, without positive cardiac involvement	Yes
Pompe–late onset Pompe disease *	Pathogenic/likely pathogenic/VUS alleles ** ≥ 2	No	Yes
Pompe–not otherwise specified *	Pathogenic/likely pathogenic/VUS alleles ** ≥ 2	No	Yes
Pompe–carrier	Pathogenic/likely pathogenic/VUS alleles = 1	No	No
Pompe–pseudodeficiency	Pseudodeficiency alleles	No	No
No disorder	No mutation found	No	No

* Regardless of the presence of pseudodeficiency allele, ** Any combination of pathogenic-pathogenic, pathogenic-likely pathogenic, pathogenic-VUS, likely pathogenic-VUS, and VUS-VUS.

We used California newborn screening data collected from 29 August 2018 through 31 August 2019. We described neonatal characteristics of all screen-positives by disease category. Demographic characteristics included newborns' sex (female, male), nursery type (Neonatal Intensive Care Unit (NICU), non-NICU), and maturity at birth (premature/<37 weeks, term/≥37 weeks). GAA enzyme activity was measured as µmol/L per hour, and the distribution of its percentage of the daily median was examined by Pompe disease categories using a box and whisker plot. We tabulated variant classification distribution across race/ethnicity groups. Race/ethnicity of each newborn was recorded as a multiple-choice check box on the GDSP Test Request Form (TRF). Single ethnic choices on the TRF were recoded to African American, Asian/Pacific Islander (API), Hispanic, non-Hispanic (NH) White, and Other. If multiple categories were reported for a newborn, we used a hierarchy to recode race/ethnicity to a single group following the order of (1) African American, (2) Hispanic, (3) API, (4) NH White, and (5) Other. Native Americans were included in the 'other' category. Variant classification information was reported for all diagnosed cases. Case notes and follow-up reports were abstracted and reviewed for the two classic IOPD patients.

All analyses were performed with SAS/STAT software version 9.4 of the SAS system for Windows (SAS Institute, Cary, NC, USA).

3. Results

3.1. Birth Prevalence

During the study period, 453,152 newborns received genetic disease screening from GDSP. Based on the GAA enzyme activity cutoff (percentage of daily median <18%), 88 newborns were screen positive for Pompe disease and received *GAA* gene sequencing to analyze mutations. Among those referred, two were diagnosed with classic IOPD, and 16 had case resolution of LOPD including 11 late-onset and five not-otherwise-specified Pompe disease, indicating an overall birth prevalence of 1 in 25,200. As of the time of this reporting, we have yet to observe a non-classic IOPD case.

Table 2 shows selected characteristics of 88 Pompe disease screen positives that have a case resolution. Male infants were more likely to be called out as screen positive. Nearly 40% of Pompe disease positive infants were in the NICU when the blood specimens were drawn, while in general, around 10% of infants were in the NICU statewide. Both classic IOPD newborns were in the NICU, and only two suspected LOPD infants needed intensive care. Interestingly, 11 out of 20 pseudodeficiency newborns and 14 of 16 false positive (no mutations found) newborns were in NICUs, suggesting other neonatal factors might play a role in reducing GAA enzyme activity in the absence of a pathogenic *GAA* gene variant. For example, eight out of 20 pseudodeficiency newborns were born prematurely.

Table 2. California Pompe disease screening results (among screen positives) by neonatal factors.

	Classic Infantile-Onset	Suspected Late-Onset	Carrier	Pseudo-Deficiency	No Disorder	Overall
Sex						
Female	1	5	14	11	6	37
Male	1	11	20	9	10	51
Nursery						
NICU	2	1	4	11	14	32
Non-NICU	0	15	30	9	2	56
Maturity						
Premature	1	2	4	8	2	17
Full term	1	14	30	12	14	71
Total	2	16	34	20	16	88
Birth prevalence	5/1,000,000	36/1,000,000	75/1,000,000	45/1,000,000		
	(1 in 226,600)	(1 in 28,300)	(1 in 13,300)	(1 in 22,700)		

3.2. GAA Activities and Pompe Disease Diagnosis

A potential link between GAA activities and forms of Pompe disease was observed. GAA values for the two patients diagnosed as classic IOPD had GAA activity significantly lower than LOPD cases and other non-disease categories (Figure 2). This observation was expected based on the pathogenesis of Pompe disease. Suspected LOPD newborns had lower GAA activity compared to carrier, pseudodeficiency, and false positive categories, although with a wide range.

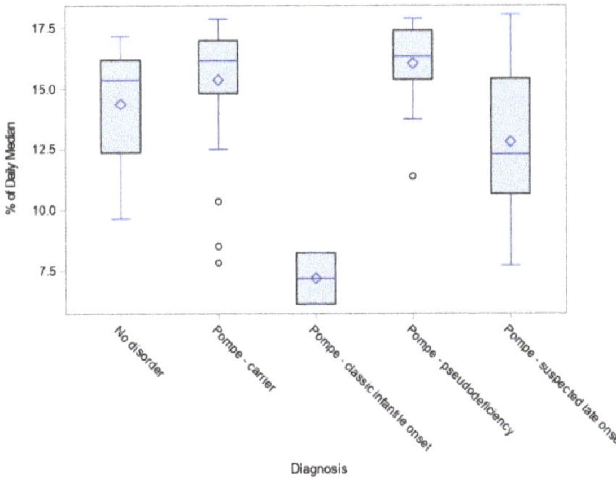

Figure 2. Acid alpha-glucosidase (GAA) activities (% of daily means) by resolution for positive Pompe disease screening.

3.3. GAA Gene Mutations and Allelic Frequency

As shown in Table 3, a total of 120 *GAA* gene variants were reported among the screen-positive cases, including 52 (43.3%) pathogenic variants, 52 (43.3%) pseudodeficiency alleles, and 16 (13.3%) VUS. The c.-32-13T>G variant was the most common pathogenic mutation (34.6% of all pathogenic variants) and was present in 10 of the 16 suspected LOPD cases, followed by c.[752C>T;761C>T]. A homozygous c.1799G>A variant was found in one of the IOPD patients; the c.1979G>A variant and the c.1754+1_1754+12delinsCCA variant were found in the other. The c.[1726G>A;2065G>A] variant was the predominant pseudodeficiency allele (80.8% of all pseudodeficiency variants).

Table 3. Pompe disease variants identified by California newborn screening.

Mutation Name	Count
Pathogenic variant	
c.-32-13T>G	18
c.[752C>T;761C>T] *	8
c.2238G>C, c.1099T>C, c.1799G>A, c.1437+1G>A, c.1548G>A, c.1579delA, c.1754+1_1754+12delinsCCA, c.1856G>A, c.1933G>A, c.1935C>A, c.1979G>A, c.2297A>C *, c.2408_2426del19, c.2560C>T, c.2646+2T>A, c.29delA, c.511del, c.546G>A, c.546G>C, c.573C>A, c.670C>T, c.925G>A	<5
Subtotal	52
Pseudodeficiency allele	
c.[1726G>A;2065G>A]	42
c.2065G>A	5
c.271G>A	5
Subtotal	52
Variant of uncertain significance	
c.1048G>A, c.1019A>G, c.1357G>A, c.1375G>A, c.1392_1393delinsTT, c.1477C>T, c.1757C>T, c.2221G>A, c.2261C>T, c.265C>T, c.266G>A **, c.316C>T, c.546+5G>T, c.726G>A, c.868A>G	<3
Subtotal	16
Total	120

* Noted as presumably non-pathogenic in the updated Pompe variant database: http://pompevariantdatabase.nl. ** Noted as presumably non-pathogenic but pathogenic with a null allele in the updated Pompe variant database: http://pompevariantdatabase.nl.

The overall pathogenic allele frequency was 115 per million (or 1 in 8700) in California's NBS population. Asian and Pacific Islander (API) and African American newborns had relatively higher frequencies (216/1,000,000 and 161/1,000,000, respectively). The overall pseudodeficiency allele frequency was also 115 per million, with API having a significantly higher rate of 432 per million (1 in 2300). Relatively higher frequencies of VUS were found in API and African American as well (Table 4).

Table 4. Allelic frequency by race/ethnicity.

Race/Ethnicity	Pathogenic		Pseudodeficiency Allele		Uncertain Significance	
	Count	Allele Frequency	Count	Allele Frequency	Count	Allele Frequency
African American (*n* = 37,340)	6	161/1,000,000 (1 in 6200)	4	107/1,000,000 (1 in 9300)	3	80/1,000,000 (1 in 12,500)
Asian/Pacific Islander (API, *n* = 69,510)	15	216/1,000,000 (1 in 4600)	30	432/1,000,000 (1 in 2300)	6	86/1,000,000 (1 in 11,600)
Hispanic (*n* = 214,049)	14	66/1,000,000 (1 in 15,300)	7	33/1,000,000 (1 in 30,600)	5	23/1,000,000 (1 in 42,800)
White (*n* = 115,281)	17	148/1,000,000 (1 in 6800)	11	95/1,000,000 (1 in 10,480)	2	17/1,000,000 (1 in 57,600)

3.4. Diagnosed Cases and Case Study of IOPD Patients

Of the 18 infants diagnosed with Pompe disease (IOPD and suspected LOPD), 12 had either a homozygous pathogenic variant or a pair of distinctive pathogenic/likely pathogenic variants (Table 5). The other six had at least one VUS, indicating a less conclusive diagnosis. Three of the 18 diagnosed cases also had a pseudodeficiency allele.

Table 5. Mutation status of diagnosed cases identified by California newborn screening.

Diagnosis	Number of Cases	Mutation Status
Pompe—classic infantile onset	1	Pathogenic, homozygous
	1	Pathogenic & pathogenic
Pompe—suspected late onset	3	Pathogenic, homozygous
	7	Pathogenic & Pathogenic/likely pathogenic
	4	Pathogenic & VUS
	1	VUS, homozygous
	1	VUS & VUS

We examined the testing results and follow-up reports on the two IOPD cases.

Case 1: This is an infant with homozygous pathogenic variant c.1799G>A, a known pathogenic mutation linked to IOPD [33,34]. The GAA confirmatory test showed "markedly reduced" enzyme activity. Further confirmatory testing showed urine glucose tetrasaccharide quantitation (Hex4) was elevated. Hypertrophic cardiomyopathy and arrhythmia were noted on the service report provided by the metabolic specialty care center clinical staff. ERT was started at two months of age.

Case 2: This is an infant with two heterozygous pathogenic variants. The c.1979G>A variant has been associated with both IOPD and LOPD [35,36]; and the c.1754+1_1754+12delinsCCA variant has no reported link to Pompe disease but was deemed as disease-causing in general [37]. Confirmatory tests found reduced GAA enzyme activity and mildly elevated Hex4. Abnormal echocardiogram and electrocardiogram results, as well as hypertrophic cardiomyopathy, were reported at the time of diagnosis. We confirmed that ERT was started but the exact starting age was unclear.

4. Discussion

The present study is one of the first reports on statewide Pompe disease screening outcomes after its placement on the RUSP, especially with a relatively large population base. California GDSP screened almost half a million babies in its first year (2018–2019) and of those referred, indicated a birth prevalence of 1 in 25,200 (IOPD and LOPD combined), which is within the range of previously reported prevalence. However, due to the rare occurrence of the disorder in the general population, only a small number of cases were reported, thus limiting the accuracy of birth prevalence calculation. With only two cases of IOPD, the birth prevalence in California (approximately 1 in 250,000) was lower than in other regions (1 in 138,000 in the Netherlands [38], 1 in 50,000 Taiwan [4,6], or 1 in 4500 in Maroon population of French Guiana [39]). However, the prevalence of potential LOPD (approximately 1 in 37,500) seems to be higher than the previously reported prevalence among the Dutch population (1 in 57,000) [38], but lower than that of Taiwan (approximately 1 in 25,000) [6]. Based on the birth prevalence of diagnosed Pompe disease cases, the calculated carrier frequency using the Hardy-Weinberg principle indicates more than five thousand carriers in our screened population. The number of carriers (34) identified from NBS was significantly fewer than that estimate because the cutoff of GAA activity in NBS aims at identifying Pompe disease cases, which have significantly lower GAA enzyme activity than that of carriers.

Six of the 16 suspected LOPD cases had at least one VUS, and since none of them have exhibited symptoms, some of their diagnoses could be changed to carrier, pseudodeficiency or no disorder based on the future clinical follow-up results. The inherent uncertainty of VUS results leads clinicians to cautiously diagnose a late-onset disorder, but affected children and their families might endure years of anxiety due to the unknown pathogenicity and consequence of the molecular findings [40]. Except when symptoms are clearly identified and a diagnosis has been made by a specialist, our observations are preliminary and incomplete given the short follow-up period of this study.

The diagnosis of Pompe disease identified by NBS is largely based on the results from molecular analysis along with supportive confirmatory testing, especially for patients who have not exhibited any symptoms. The high occurrence of the pathogenic c.-32-13T>G variant in our screen positive samples (40.4% of all pathogenic variants) echoed findings from literature, which reported an allelic frequency from 40% to 70% [41]. For newly screened rare disorders with late-onset phenotypes like Pompe disease, one of the greatest challenges for screening is the VUS category in which cases have an unknown pathogenic molecular profile. Some VUS may eventually be recognized as pathogenic, but barriers to receiving a thorough clinical work-up or ongoing clinical follow-up (such as factors associated with access to care), could play a role in obtaining a more definitive diagnosis later. With a more developed global registry and variant database [32,34] future screening could yield more predictive results.

California has a vastly diverse population. In our study, Pompe disease-positive newborns with Asian and Pacific Islander (API) ancestry had a high occurrence of pseudodeficiency alleles, especially the c.[1726G>A;2065G>A] variant, which represents 80% of all the pseudodeficiency mutations. This finding confirmed the results from other studies with Asian populations [22,42,43]. Unlike these other studies, we did not find any Pompe disease cases (IOPD or LOPD) among nearly 70,000 API newborns, and we only found one c.1935C>A (linked to c.[1726G>A;2065G>A]), which was identified as the most common pathogenic *GAA* variants among Asian countries. African American newborns had a birth prevalence of 54 per million (or 1 in 18,700), which was the highest among all groups. This result may be indicative of a potentially high Pompe disease birth prevalence among African Americans, but more data are needed to be conclusive. Previous research identified c.2560C>T as the most common *GAA* variant among African Americans [34,44]. We did not have a large enough sample size (*n* = 7) of African American infants who had variants to confirm the finding. The only c.2560C>T variant, however, was indeed detected in an African American sample.

In most of the study period (before 21 August 2019), every newborn with GAA activity ≤ 18% of the daily median was flagged as an urgent call-out by the laboratory before the results of *GAA* gene sequencing was available. About six months after the Pompe disease newborn screening began, NBS received communications from clinical specialists about the follow-up burden for both patients and providers due to the large number of patients being referred; many of them were either pseudodeficiency or no mutation based on the sequencing findings. Although previous research showed that MS/MS analysis of GAA activity could separate pseudodeficiency and Pompe disease cases [45], our screening test results still showed some overlap in GAA activity for these two groups. GDSP evaluated the available data and modified the protocol in August 2019 to flag only the cases with GAA activity less than 10% of the daily median for urgent call-out (the two IOPD cases identified through the program were well-below this threshold). Newborns with GAA activities between 10% and 18% of the daily median are only referred if the molecular results show pathogenic, likely pathogenic, or VUS mutations. In other words, we wait so that screen-positive newborns with homozygous or heterozygous pseudodeficiency alleles or no mutations are not referred to the specialty care centers for further follow-up. This serves as a good example of how synergy between providers and the newborn screening program minimized the unnecessary referrals and improved screening performance. If we later find infants who have GAA activities between 10% and 18% of the daily median diagnosed with IOPD, we will consider adjusting the cutoff again for urgent call-outs.

More than four years after Pompe disease was added to the RUSP, the adoption and implementation of newborn screening at the state level has been at a moderate pace. In the first year of Pompe disease newborn screening in California, we have gained a better understanding of the disease distribution at the population level, and most importantly, now have experience and evidence to support effective screening. With a robust long-term follow-up component, GDSP values the necessity of monitoring all potential cases, including those with a VUS [46]. The growing knowledge from long-term follow-up will further improve our understanding of the clinical significance of these cases, especially when case management algorithms are still undeveloped for asymptomatic patients [47].

While the two newborns with IOPD were identified while in the NICU, almost all of the newborns with LOPD were identified in the regular nursery. These newborns were asymptomatic and unlikely to be identified as at risk for Pompe disease except by screening. Now that treatment is warranted before symptoms develop, the value of population-based screening is clear: to identify the youngest candidates for treatments that can reduce life-long disability [48].

Author Contributions: Conceptualization, R.S.O., H.T., L.F., and S.S.; methodology, H.T. and S.S.; formal analysis, H.T. and J.M.; resources, T.B.; Data curation, H.T., J.M., and D.M.; writing—original draft preparation, H.T.; writing—review and editing, L.F., S.S., J.M., and D.M.; visualization, H.T., J.M., and S.S.; supervision, L.F. and R.S.O.; project administration, T.B. All authors have read and agreed to the published version of the manuscript.

Funding: This research received no external funding.

Acknowledgments: We want to extend our thanks to the staff at the Genetic Disease Laboratory who conduct Pompe disease testing and interpretation of test results as well as to the staff at Perkin-Elmer Life Sciences who conduct the genomic sequencing of screen-positive cases. We also want to thank the staff at the state-contracted California Metabolic Special Care Centers who provide follow-up diagnostic services for referred newborns and the data for our ongoing program evaluation efforts.

Conflicts of Interest: The authors declare no conflict of interest.

References

1. Reuser, A.J.J.; Hirschhorn, R.; Kroos, M.A. Pompe Disease: Glycogen storage disease type II: Acid α-glucosidase (acid maltase) deficiency. In *The Online Metabolic and Molecular Bases of Inherited Disease*; Valle, D., Antonarakis, S., Ballabio, A., Eds.; McGraw-Hill: New York, NY, USA, 2018.

2. Martiniuk, F.; Chen, A.; Mack, A.; Arvanitopoulos, E.; Chen, Y.; Rom, W.N.; Codd, W.J.; Hanna, B.; Alcabes, P.; Raben, N.; et al. Carrier frequency for glycogen storage disease type II in New York and estimates of affected individuals born with the disease. *Am. J. Med. Genet.* **1998**, *79*, 69–72. [CrossRef]

3. Bashan, N.; Potashnik, R.; Barash, V.; Gutman, A.; Moses, S.W. Glycogen storage disease type II in Israel. *Isr. J. Med. Sci.* **1988**, *24*, 224–227.

4. Chiang, S.C.; Hwu, W.L.; Lee, N.C.; Hsu, L.W.; Chien, Y.H. Algorithm for Pompe disease newborn screening: Results from the Taiwan screening program. *Mol. Genet. Metab.* **2012**, *106*, 281–286. [CrossRef]

5. Scott, C.R.; Elliott, S.; Buroker, N.; Thomas, L.I.; Keutzer, J.; Glass, M.; Gelb, M.H.; Turecek, F. Identification of infants at risk for developing Fabry, Pompe, or mucopolysaccharidosis-I from newborn blood spots by tandem mass spectrometry. *J. Pediatr.* **2013**, *163*, 498–503. [CrossRef]

6. Chien, Y.H.; Lee, N.C.; Huang, H.J.; Thurberg, B.L.; Tsai, F.J.; Hwu, W.L. Later-onset Pompe disease: Early detection and early treatment initiation enabled by newborn screening. *J. Pediatr.* **2011**, *158*, 1023–1027. [CrossRef]

7. Dasouki, M.; Jawdat, O.; Almadhoun, O.; Pasnoor, M.; McVey, A.L.; Abuzinadah, A.; Herbelin, L.; Barohn, R.J.; Dimachkie, M.M. Pompe disease: Literature review and case series. *Neurol. Clin.* **2014**, *32*, 751–776. [CrossRef]

8. Van den Hout, H.M.; Hop, W.; van Diggelen, O.P.; Smeitink, J.A.; Smit, G.P.; Poll-The, B.T.; Bakker, H.D.; Loonen, M.C.; de Klerk, J.B.; Reuser, A.J.; et al. The natural course of infantile Pompe's disease: 20 original cases compared with 133 cases from the literature. *Pediatrics* **2003**, *112*, 332–340. [CrossRef]

9. Umapathysivam, K.; Hopwood, J.J.; Meikle, P.J. Correlation of acid alpha-glucosidase and glycogen content in skin fibroblasts with age of onset in Pompe disease. *Clin. Chim. Acta* **2005**, *361*, 191–198. [CrossRef]

10. Chien, Y.H.; Hwu, W.L.; Lee, N.C. Pompe disease: Early diagnosis and early treatment make a difference. *Pediatr. Neonatol.* **2013**, *54*, 219–227. [CrossRef]

11. Lai, C.J.; Hsu, T.R.; Yang, C.F.; Chen, S.J.; Chuang, Y.C.; Niu, D.M. Cognitive development in infantile-Onset Pompe disease under very early enzyme replacement therapy. *J. Child. Neurol.* **2016**, *31*, 1617–1621. [CrossRef]

12. Tarnopolsky, M.; Katzberg, H.; Petrof, B.J.; Sirrs, S.; Sarnat, H.B.; Myers, K.; Dupre, N.; Dodig, D.; Genge, A.; Venance, S.L.; et al. Pompe disease: Diagnosis and management. Evidence-based guidelines from a Canadian expert panel. *Can. J. Neurol. Sci.* **2016**, *43*, 472–485. [CrossRef] [PubMed]

13. Kronn, D.F.; Day-Salvatore, D.; Hwu, W.L.; Jones, S.A.; Nakamura, K.; Okuyama, T.; Swoboda, K.J.; Kishnani, P.S. Pompe Disease Newborn Screening Working Group. Management of confirmed newborn-Screened patients with Pompe disease across the disease spectrum. *Pediatrics* **2017**, *140*, S24–S45.

14. Ortolano, R.; Baronio, F.; Masetti, R.; Prete, A.; Cassio, A.; Pession, A. Letter to the Editors: Concerning "Divergent clinical outcomes of alphaglucosidase enzyme replacement therapy in two siblings with infantile-onset Pompe disease treated in the symptomatic or pre-symptomatic state" by Takashi M et al. *Mol. Genet. Metab. Rep.* **2017**, *11*, 1. [CrossRef] [PubMed]

15. Umapathysivam, K.; Whittle, A.M.; Ranieri, E.; Bindloss, C.; Ravenscroft, E.M.; van Diggelen, O.P.; Hopwood, J.J.; Meikle, P.J. Determination of acid alpha-Glucosidase protein: Evaluation as a screening marker for Pompe disease and other lysosomal storage disorders. *Clin. Chem.* **2000**, *46*, 1318–1325. [CrossRef] [PubMed]

16. Gelb, M.H.; Turecek, F.; Scott, C.R.; Chamoles, N.A. Direct multiplex assay of enzymes in dried blood spots by tandem mass spectrometry for the newborn screening of lysosomal storage disorders. *J. Inherit. Metab. Dis.* **2006**, *29*, 397–404. [CrossRef] [PubMed]

17. Spacil, Z.; Tatipaka, H.; Barcenas, M.; Scott, C.R.; Turecek, F.; Gelb, M.H. High-Throughput assay of 9 lysosomal enzymes for newborn screening. *Clin. Chem.* **2013**, *59*, 502–511. [CrossRef]

18. Gucciardi, A.; Legnini, E.; Di Gangi, I.M.; Corbetta, C.; Tomanin, R.; Scarpa, M.; Giordano, G. A column-Switching HPLC-MS/MS method for mucopolysaccharidosis type I analysis in a multiplex assay for the simultaneous newborn screening of six lysosomal storage disorders. *Biomed. Chromatogr.* **2014**, *28*, 1131–1139. [CrossRef]

19. Tortorelli, S.; Turgeon, C.T.; Gavrilov, D.K.; Oglesbee, D.; Raymond, K.M.; Rinaldo, P.; Matern, D. Simultaneous testing for 6 lysosomal storage disorders and x-Adrenoleukodystrophy in dried blood spots by tandem mass spectrometry. *Clin. Chem.* **2016**, *62*, 1248–1254. [CrossRef]

20. Chien, Y.H.; Chiang, S.C.; Zhang, X.K.; Keutzer, J.; Lee, N.C.; Huang, A.C.; Chen, C.A.; Wu, M.H.; Huang, P.H.; Tsai, F.J.; et al. Early detection of Pompe disease by newborn screening is feasible: Results from the Taiwan screening program. *Pediatrics* **2008**, *122*, e39–e45. [CrossRef]

21. Burton, B.K. Newborn screening for Pompe disease: An update, 2011. *Am. J. Med. Genet. C. Semin. Med. Genet.* **2012**, *160C*, 8–12. [CrossRef]

22. Momosaki, K.; Kido, J.; Yoshida, S.; Sugawara, K.; Miyamoto, T.; Inoue, T.; Okumiya, T.; Matsumoto, S.; Endo, F.; Hirose, S.; et al. Newborn screening for Pompe disease in Japan: Report and literature review of mutations in the GAA gene in Japanese and Asian patients. *J. Hum. Genet.* **2019**, *64*, 741–755. [CrossRef]

23. Wittmann, J.; Karg, E.; Turi, S.; Legnini, E.; Wittmann, G.; Giese, A.K.; Lukas, J.; Golnitz, U.; Klingenhager, M.; Bodamer, O.; et al. Newborn screening for lysosomal storage disorders in hungary. *JIMD Rep.* **2012**, *6*, 117–125. [PubMed]

24. Kemper, A.R. Condition Review Workgroup Evidence Report: Newborn Screening for Pompe Disease. Available online: https://www.hrsa.gov/sites/default/files/hrsa/advisory-committees/heritable-disorders/rusp/previous-nominations/pompe-external-evidence-review-report-2013.pdf (accessed on 14 November 2019).

25. Advisory Committee on Heritable Disorders in Newborns and Children. Recommended Uniform Screening Panel [As of July 2018]. Available online: http://www.hrsa.gov/advisorycommittees/mchbadvisory/heritabledisorders/recommendedpanel (accessed on 14 November 2019).

26. The Newborn Screening Technical Assistance and Evaluation Program (NewSTEPs), Association of Public Health Laboratories. Disorders Screening Status Map: Pompe. Available online: https://www.newsteps.org/resources/newborn-screening-status-all-disorders (accessed on 14 November 2019).

27. California State Legislature. Senate Bill No.1095: Newborn Screening Program. 2016. Available online: https://leginfo.legislature.ca.gov/faces/billTextClient.xhtml?bill_id=201520160SB1095 (accessed on 14 November 2019).

28. Califorina Health and Safety Code, Sections 125001, amended 2016. Available online: https://leginfo.legislature.ca.gov/faces/codes_displayText.xhtml?lawCode=HSC&division=106.&title=&part=5.&chapter=1.&article=2 (accessed on 19 December 2019).

29. Bronstein, M.G.; Pan, R.J.; Dant, M.; Lubin, B. Leveraging evidence-Based public policy and advocacy to advance newborn screening in California. *Pediatrics* **2019**, *143*, e20181886. [CrossRef] [PubMed]

30. The Newborn Screening Technical Assistance and Evaluation Program (NewSTEPs), Association of Public Health Laboratories. NewSTEPs Data Repository Case Definition. Available online: https://www.newsteps.org/case-definitions (accessed on 21 November 2019).

31. Richards, S.; Aziz, N.; Bale, S.; Bick, D.; Das, S.; Gastier-Foster, J.; Grody, W.W.; Hegde, M.; Lyon, E.; Spector, E.; et al. Standards and guidelines for the interpretation of sequence variants: A joint consensus recommendationof the American College of Medical Genetics and Genomics and the Association for Molecular Pathology. *Genet. Med.* **2015**, *17*, 405–424. [CrossRef] [PubMed]

32. Nino, M.Y.; in't Groen, S.L.M.; Bergsma, A.J.; van der Beek, N.A.M.E.; Kroos, M.; Hoogeveen-Westerveld, M.; van der Ploeg, A.T.; Pijnappel, W.W.M.P. Extension of the Pompe mutation database by linking disease-Associated variants to clinical severity. *Hum. Mutat.* **2019**, *40*, 1954–1967. [CrossRef] [PubMed]

33. Reuser, A.J.J.; van der Ploeg, A.T.; Chien, Y.H.; Llerena, J., Jr.; Abbott, M.A.; Clemens, P.R.; Kimonis, V.E.; Leslie, N.; Maruti, S.S.; Sanson, B.J.; et al. GAA variants and phenotypes among 1,079 patients with Pompe disease: Data from the Pompe Registry. *Hum. Mut.* **2019**, *40*, 2146–2164. [CrossRef]

34. Herzog, A.; Hartung, R.; Reuser, A.J.; Hermanns, P.; Runz, H.; Karabul, N.; Gokce, S.; Pohlenz, J.; Kampmann, C.; Lampe, C.; et al. A cross-Sectional single-centre study on the spectrum of Pompe disease, German patients: Molecular analysis of the GAA gene, manifestation and genotype-phenotype correlations. *Orphanet J. Rare Dis.* **2012**, *7*, 35. [CrossRef]

35. Burrow, T.A.; Bailey, L.A.; Kinnett, D.G.; Hopkin, R.J. Acute progression of neuromuscular findings in infantile Pompe disease. *Pediatr. Neurol.* **2010**, *42*, 455–458. [CrossRef]

36. Nazari, F.; Sinaei, F.; Nilipour, Y.; Fatehi, F.; Streubel, B.; Ashrafi, M.R.; Aryani, O.; Nafissi, S. Late-Onset pompe disease in Iran: A clinical and genetic report. *Muscle Nerve* **2017**, *55*, 835–840. [CrossRef]

37. Lek, M.; Karczewski, K.J.; Minikel, E.V.; Samocha, K.E.; Banks, E.; Fennell, T.; O'Donnell-Luria, A.H.; Ware, J.S.; Hill, A.J.; Cummings, B.B.; et al. Analysis of protein-Coding genetic variation in 60,706 humans. *Nature* **2016**, *536*, 285–291. [CrossRef]

38. Ausems, M.G.; Verbiest, J.; Hermans, M.P.; Kroos, M.A.; Beemer, F.A.; Wokke, J.H.; Sandkuijl, L.A.; Reuser, A.J.; van der Ploeg, A.T. Frequency of glycogen storage disease type II in The Netherlands: Implications for diagnosis and genetic counselling. *Eur. J. Hum. Genet.* **1999**, *7*, 713–716. [CrossRef]

39. Elenga, N.; Verloes, A.; Mrsic, Y.; Basurko, C.; Schaub, R.; Cuadro-Alvarez, E.; Kom-Tchameni, R.; Carles, G.; Lambert, V.; Boukhari, R.; et al. Incidence of infantile Pompe disease in the Maroon population of French Guiana. *BMJ Paediatr. Open* **2018**, *2*, e000182. [CrossRef] [PubMed]

40. Pruniski, B.; Lisi, E.; Ali, N. Newborn screening for Pompe disease: Impact on families. *J. Inherit. Metab. Dis.* **2018**, *41*, 1189–1203. [CrossRef] [PubMed]

41. Peruzzo, P.; Pavan, E.; Dardis, A. Molecular genetics of Pompe disease: A comprehensive overview. *Ann. Transl. Med.* **2019**, *7*, 278. [CrossRef] [PubMed]

42. Kroos, M.A.; Mullaart, R.A.; Van Vliet, L.; Pomponio, R.J.; Amartino, H.; Kolodny, E.H.; Pastores, G.M.; Wevers, R.A.; Van der Ploeg, A.T.; Halley, D.J.; et al. p.[G576S.; E689K]: Pathogenic combination or polymorphism in Pompe disease? *Eur. J. Hum. Genet.* **2008**, *16*, 875–879. [CrossRef] [PubMed]

43. Lee, J.H.; Shin, J.H.; Park, H.J.; Kim, S.Z.; Jeon, Y.M.; Kim, H.K.; Kim, D.S.; Choi, Y.C. Targeted population screening of late onset Pompe disease in unspecified myopathy patients for Korean population. *Neuromuscul. Disord.* **2017**, *27*, 550–556. [CrossRef]

44. Becker, J.A.; Vlach, J.; Raben, N.; Nagaraju, K.; Adams, E.M.; Hermans, M.M.; Reuser, A.J.; Brooks, S.S.; Tifft, C.J.; Hirschhorn, R.; et al. The African origin of the common mutatoin in African American patients with glycogen-Storage disease type II. *Am. J. Hum. Genet.* **1998**, *62*, 991–994. [CrossRef]

45. Liao, H.C.; Chan, M.J.; Yang, C.F.; Chiang, C.C.; Niu, D.M.; Huang, C.K.; Gelb, M.H. Mass spectrometry but not fluorimetry distinguisheds affected and pseudodeficiencies in newborn screening for Pompe disease. *Clin. Chem.* **2017**, *63*, 1271–1277. [CrossRef]

46. Feuchtbaum, L.; Yang, J.; Currier, R. Follow-Up status during the first 5 years of life for metabolic disorders on the federal Recommended Uniform Screening Panel. *Genet. Med.* **2018**, *20*, 831–839. [CrossRef]

47. Kemper, A.R.; Boyle, C.A.; Brosco, J.P.; Grosse, S.D. Ensuring the life-Span benefits of newborn screening. *Pediatrics* **2019**, *144*, e20190904. [CrossRef]
48. Yang, C.F.; Yang, C.C.; Liao, H.C.; Huang, L.Y.; Chiang, C.C.; Ho, H.C.; Lai, C.J.; Chu, T.H.; Yang, T.F.; Hsu, T.R.; et al. Very early treatment for infantile-Onset Pompe disease contributes to better outcomes. *J. Pediatr.* **2016**, *169*, 174–180. [CrossRef]

International Journal of
Neonatal Screening

Article

Newborn Screening for Pompe Disease: Pennsylvania Experience

Can Ficicioglu [1,*], Rebecca C. Ahrens-Nicklas [1], Joshua Barch [2], Sanmati R. Cuddapah [1],
Brenda S. DiBoscio [1], James C. DiPerna [3], Patricia L. Gordon [4], Nadene Henderson [2],
Caitlin Menello [1], Nicole Luongo [1], Damara Ortiz [2] and Rui Xiao [5,6]

[1] Division of Human Genetics/Metabolism, Children's Hospital of Philadelphia, Perelman School of Medicine
    at the University of Pennsylvania, Philadelphia, PA 19104, USA;
    ahrensNicklasR@email.chop.edu (R.C.A.-N.); cuddapahs@email.chop.edu (S.R.C.);
    dibosciob1@email.chop.edu (B.S.D.); menelloc@email.chop.edu (C.M.); Luongon@email.chop.edu (N.L.)
[2] Department of Pediatrics, Division of Medical Genetics, UPMC Children's Hospital of Pittsburgh, Pittsburgh,
    PA 15224, USA; joshua.barch@chp.edu (J.B.); nadene.henderson@chp.edu (N.H.);
    damara.ortiz@chp.edu (D.O.)
[3] PerkinElmer, Mass Spectroscopy Unit, Pittsburgh, PA 15275, USA; James.DiPerna@perkinelmer.com
[4] Division of Human Genetics, Penn State Heath Children's Hospital, Penn State University College of
    Medicine, Hershey, PA 17033, USA; pgordon@pennstatehealth.psu.edu
[5] Department of Pediatrics, Division of Biostatistics, Children's Hospital of Philadelphia, Philadelphia,
    PA 19104, USA; xiaor@email.chop.edu
[6] Department of Biostatistics, Epidemiology & Informatics, Perelman School of Medicine at the University
    of Pennsylvania, Philadelphia, PA 19104, USA
* Correspondence: ficicioglu@email.chop.edu

Received: 2 October 2020; Accepted: 6 November 2020; Published: 13 November 2020

Abstract: Pennsylvania started newborn screening for Pompe disease in February 2016. Between February 2016 and December 2019, 531,139 newborns were screened. Alpha-Glucosidase (GAA) enzyme activity is measured by flow-injection tandem mass spectrometry (FIA/MS/MS) and full sequencing of the GAA gene is performed as a second-tier test in all newborns with low GAA enzyme activity [<2.10 micromole/L/h]. A total of 115 newborns had low GAA enzyme activity and abnormal genetic testing and were referred to metabolic centers. Two newborns were diagnosed with Infantile Onset Pompe Disease (IOPD), and 31 newborns were confirmed to have Late Onset Pompe Disease (LOPD). The incidence of IOPD + LOPD was 1:16,095. A total of 30 patients were compound heterozygous for one pathogenic and one variant of unknown significance (VUS) mutation or two VUS mutations and were defined as suspected LOPD. The incidence of IOPD + LOPD + suspected LOPD was 1: 8431 in PA. We also found 35 carriers, 15 pseudodeficiency carriers, and 2 false positive newborns.

Keywords: Pompe disease; newborn screening; alpha glucosidase

---

1. Introduction

Pompe disease is an autosomal recessive inborn error of metabolism caused by mutations in the glucosidase alpha acid (*GAA*) gene located on the long arm of chromosome 17q25.2-q25.3. These genetic mutations cause deficient GAA enzyme activity [1]. Due to the deficiency of this enzyme, glycogen cannot be metabolized in lysosomes. This causes glycogen accumulation that damages cells throughout the body, especially muscle cells. Pathologic changes in muscle usually begin long before patients present with symptoms [2].

The previously estimated overall incidence of Pompe disease is 1 in 40,000 [1 in 138,000 for Infantile Onset Pompe Disease (IOPD) and 1 in 57,000 for Late Onset Pompe Disease (LOPD)]

in the Netherlands [3]. This roughly corresponds to the incidence of clinically identified cases in New York state [4]. The frequency appears to vary significantly in different ethnic groups from 1 in 14,000 to 1 in 600,000 [1,3–7]. Recorded incidence is highest in African Americans and lowest in Portuguese populations.

Patients with IOPD present with hypertrophic cardiomyopathy, hypotonia, macroglossia, feeding difficulties, and failure to thrive at around 2 months of age [8]. Some cases develop hypertrophic cardiomyopathy in utero. IOPD is rapidly progressive, and if left untreated, patients usually die in the first year of life. IOPD patients, diagnosed clinically or by newborn screening, always have elevated CK levels as a marker of muscle damage, and elevated urinary glucose tetrasaccharide (Glc4 or Hex4). Hex4 was shown to correlate with glycogen content in quadriceps biopsies in patients with IOPD [9]. Hex 4 is not only useful in the diagnosis of Pompe but also in monitoring the response to ERT [10].

Patients with LOPD present later in life with proximal muscle weakness, gait abnormalities, respiratory insufficiency, poor weight gain, and swallowing difficulties [11]. Individuals with LOPD do not typically develop hypertrophic cardiomyopathy, but some experience arrhythmias. LOPD patients who are diagnosed clinically often, but not always, have elevated CK and Hex 4 levels. Baseline evaluation of Hex 4 levels in LOPD cases detected through NBS have been reported to be within normal range [10].

Diagnostic Odyssey: Before implementation of newborn screening, there was, on average, a 3 month-delay in diagnosing IOPD after the onset of symptoms [12]. In LOPD, symptoms may begin anytime from infancy to adulthood. In pediatric onset cases, symptom onset occurs at an average of 6 years of age, yet the diagnosis is made at an average of 18 years of age. Therefore, on average, there is a 12-year delay in diagnosis [12]. The average age of symptom onset in adult-onset LOPD is 35 years of age, with a 7-year delay in diagnosis after symptom onset [12]. Non-specific symptoms, seen in other more common disorders, coupled with lack of knowledge about this rare disorder usually result in significant diagnostic delays. These factors importantly shape an individual patient's odyssey and outcomes with all forms of Pompe disease.

Enzyme replacement therapy (ERT) has been available since 2006 for all forms of Pompe disease. It has dramatically changed patient outcomes. [13]. ERT improves cardiac and skeletal symptoms and slows down the progression of disease. Patients with severe disease, who cannot produce any natural GAA enzyme, are classified as Cross-Reactive Immunological Material (CRIM) negative. In such CRIM negative patients, the immune system identifies the administered enzyme as foreign and produces antibodies that make the enzyme therapy less effective [14]. Immune modulation therapy can be used to inhibit the development of antibody response in CRIM negative patients at the time of the start of their ERT [15–17].

This life-changing therapy is more effective when initiated before the onset of symptoms [18–22]. When we consider that (a) pathology begins long before patients present with symptoms and (b) many patients have a protracted diagnostic odyssey, shortening the diagnostic delay through newborn screening and beginning treatment as soon as possible is crucial for all forms of Pompe disease.

A pilot newborn screening (NBS) was launched in Taiwan in 2005. Chien et al. demonstrated the importance of NBS not only for IOPD but also for LOPD [19–21]. Between 2005 and 2009, 344,056 newborns were screened in Taiwan, and 13 cases of LOPD were detected [21]. A total of 4 of 13 patients were put on ERT because of hypotonia, muscle weakness, delayed developmental milestones/motor skills, or elevated CK levels starting at the ages of 1.5, 14, 34, and 36 months. Muscle biopsy specimens obtained from the treated patients revealed increased storage of glycogen and lipids [21]. Soon after Taiwan, several pilot programs in Italy, Australia, Japan, Korea, USA, and Hungary tested the feasibility of Pompe NBS and reported the incidence of and the impact of the disease [23].

In the USA, Pompe disease was added to the recommended universal newborn screening panel (RUSP) in February 2015 [24]. Some states such as Missouri, Illinois, and New York started screening for Pompe before the RUSP recommendation. Pennsylvania started Pompe screening in February 2016.

In this paper, we present Pennsylvania data and experience on Pompe NBS. Our aim is three-fold: to augment state-based knowledge of the disease and its diagnosis through NBS; to document the benefits and challenges of NBS for Pompe disease in Pennsylvania; and to encourage the speedy adoption of NBS for Pompe by other states and countries.

2. Materials and Methods

In Pennsylvania, NBS testing is done by PerkinElmer newborn screening laboratory in Pittsburgh, PA. GAA enzyme activity is measured by flow-injection tandem mass spectrometry (FIA/MS/MS) using the algorithm shown in Figure 1 [25]. Blood is obtained at around 36 h of life. A second sample is requested if the first GAA enzyme activity is below the cutoff value. If this second GAA enzyme activity is below the cutoff, the second-tier test, *GAA* full gene sequencing (Next Generation Sequencing), is performed. In Pennsylvania, newborn screening is free, and both GAA enzyme level and full gene sequencing are done as part of Pompe NBS without any charges. The cutoff was established at ~18% of the apparently normal newborn mean GAA activity and is based on the analysis of ~1000 apparently normal newborns as well as 9 confirmed positive Pompe patients. In PerkinElmer laboratory, it was determined that the daily patient means are stable, and thus, the use of a fixed cutoff has been utilized and has been shown to be effective. The use of fixed cutoffs also makes the daily evaluation of data much less complex. Results falling below this cutoff are considered abnormal.

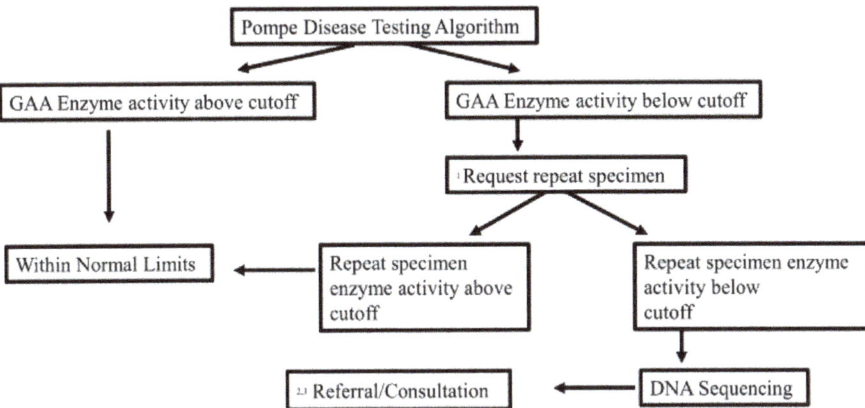

Figure 1. Flow chart of screening algorithm. [1] Pompe Repeat Request: Results show decreased enzyme activity for lysosomal alpha-glucosidase (GAA). We recommend a repeat dried filter paper blood specimen within 48 hours after the reporting of the first abnormal test. You should discuss the case with one of the metabolism referral centers if newborn has cardiomyopathy or symptoms of Pompe disease such as hypotonia and feeding issues. [2] Pompe Referral (Sequencing positive): Results continue to show decreased enzyme activity for Lysosomal alpha-Glucosidase (GAA). This result may be associated with Pompe disease. We recommend referral to a metabolic specialist. [3] Pompe Consult (Sequencing pseudodeficiency or no variant): Results continue to show decreased enzyme activity for Lysosomal alpha-Glucosidase (GAA). Consultation with a metabolic specialist may be considered to review and interpret these results in context with the patient's clinical presentation.

A single 1/8-inch (3mm) Dried Blood Spot (DBS) disc is extracted with 30 µL of cocktail solution containing β-Glucocerebrosidase (ABG), Acid Sphingomyelinase (ASM), Alpha-Glucosidase (GAA), Alpha-Galactosidase (GLA), Galactocerebrosidase (GALC) and Alpha-Iduronidase (IDUA) substrates and internal standards (in a buffer of pH 4.8) at 37 °C for 18 h +/− 2 h. The enzymatic reactions are quenched with 50% ethyl acetate: methanol followed by a liquid–liquid extraction using 50% ethyl acetate: water.

An aliquot of the organic top layer is transferred into a clean deep well plate, dried under a gentle stream of nitrogen gas, reconstituted in 80%/20%/0.2% acetonitrile/water/formic acid solution, and subjected to flow-injection tandem mass spectrometry (FIA/MS/MS). The electrospray source is operated in positive mode, and the analytes are interrogated in multiple reaction monitoring (MRM) mode. Blank filter paper spots are analyzed for background correction.

Enzyme activities (in units of μmol/L/h) are defined as the amount of substrate hydrolyzed by the enzyme in the reaction and are determined by calculating the ion abundance ratio of product to internal standard, multiplied by the volume and concentration of internal standard, divided by the response factor ratio of product to internal standard, sample incubation time, and sample blood volume (Enzyme activity = $(P/IS)*[IS]*V_{IS}/RF/3.1$ μL/time). A sample volume of 3.1 μL is assigned to a 1/8-inch DBS punch. A fixed GAA enzyme activity cutoff of 2.10 micromole/L/h is utilized. A repeat specimen is requested on an initial GAA enzyme result of <2.10 μmol/L/h. A repeat specimen with a second GAA enzyme level of <2.10 μmol/L/h is reflexed to full *GAA* gene sequencing. Turn-around-time for GAA enzyme testing is <72 h from specimen receipt. Full *GAA* sequencing turn-around time is 7–10 days from reflex.

Newborns with low enzyme activity (<2.10 μmol/L/h) and at least one variant [pathogenic, likely pathogenic, or variant of unknown significance (VUS)] are referred to one of the seven metabolic referral centers in Pennsylvania by pediatricians and department of health newborn screening program. Newborns with low enzyme activity and sequencing revealing only pseudodeficiency allele(s) are not referred to metabolic centers by the state. However, some pediatricians refer them to the metabolic centers for further clarification.

After the initial metabolic evaluation and confirmatory test results, patients are classified as IOPD, LOPD, "suspected" LOPD, or carrier. If patients carry one pathogenic variant and one VUS or two VUS, they are defined as "suspected" LOPD. After genetic counseling, carriers and those with pseudo-deficiency mutations are discharged with no further follow-up; all others are followed by a metabolic geneticist. Although there are slight differences among Pennsylvania centers in the initial work up, confirmatory GAA enzyme activity, aspartate aminotransferase (AST), alanine aminotransferase (ALT), creatine kinase (CK), urine glucose tetrasaccharide (Glc4 or Hex4) levels are tested in all cases. At the outset of NBS for Pompe, all centers performed electrocardiogram (ECG) and echocardiogram (ECHO) at the time of referral. As centers gained more experience, some decided not to perform ECG and ECHO at the first visit in cases with late onset mutations. Patients with LOPD and suspected LOPD are usually seen every 3 months in the first year and subsequently, every 6 months. In IOPD patients, treatment, including appropriate immunomodulation, is initiated immediately.

Statistical Analysis: The demographic and laboratory data were summarized by median and interquartile range (IQR) for continuous variables and frequency and percentage for categorical variables. Wilcoxon rank sum test was used to compare the GAA enzyme activity, as well as AST, ALT, CK, Hex4 levels between groups, either disease groups or genotype groups. Two-sided *p* values less than 0.05 were considered statistically significant. All data analyses were performed using SAS 9.4.

The Institutional Review Board reviewed and deemed this study exempt (IRB 20-017914).

3. Results

Between February 2016 and December 2019, 531,139 newborns were screened for Pompe disease. A total of 180 newborns (0.03%) had a low GAA enzyme activity on the first NBS. Repeat GAA enzyme activity from a second filter paper specimen was lower than the cutoff in 115 newborns (0.02%), and subsequently full *GAA* gene sequencing was performed. The screening algorithm is shown in Figure 1. Final diagnosis, screening and biochemical data from patients with a confirmed diagnosis are summarized in Tables 1–3. All positive cases were referred to one of the seven metabolic centers. The incidence of IOPD + LOPD was 1:16,095 and IOPD + LOPD + suspected LOPD was 1:8431 in PA.

Based on the PA DOH records, 5 newborns with positive screening were lost to follow up and never seen in metabolic centers. All 5 newborns were carriers based on the genetic test result.

Table 1. Final diagnosis of newborn screening for Pompe disease.

Pompe Disease	*n*	Incidence
IOPD	2	1:265,570
LOPD	31	1:17,134
Suspected LOPD	30	1:17,705
IOPD+LOPD	33	1:16,095
IOPD+ LOPD+ suspected LOPD	63	1:8431
Carriers	35	1:15,175
Pseudodeficiency	15	1:35,409
False positive (Normal confirmatory GAA activity, negative GAA full gene sequencing)	2	1:265,570

Total newborns screened: 531,139, Total positive *n* (%):115 (0.02).

IOPD: Two patients were diagnosed with IOPD. Their GAA enzyme levels were both 0.65 micromole/L/h at the first screening and 0.22 and 0.34 at the second screening, respectively. They both had lower GAA enzyme activities compared to carriers and pseudodeficiency; yet, there was overlap with other forms of Pompe disease. Both patients were compound heterozygous for mutations previously reported in patients with IOPD (Table 2).

The first patient was in the hospital and admitted for hypertrophic cardiomyopathy when the NBS was reported. He had elevated CK, AST, ALT, BNP, and Hex4 levels. ECG showed biventricular hypertrophy. ECHO findings included (1) moderately to severely decreased left ventricular systolic wall motion, (2) moderate hypertrophy of the left ventricle, and (3) severe right ventricular hypertrophy. The CRIM status was determined based on the genotype. ERT with immunomodulation (CRIM negative protocol) was started at 21 days of age [16]. At 31 months of age, he has normal cardiac functions but global developmental delay. He has obstructive sleep apnea and requires BiPAP at night. He continues to receive physical, occupational, and speech therapies.

The second patient had elevated CK, AST, ALT, BNP, and Hex4 levels. ECG showed short PR interval and biventricular hypertrophy. ECHO findings were (1) severe septal left ventricular hypertrophy and (2) mildly hypertrophied right ventricle. He was CRIM positive based on his genotype. ERT with immuno-modulation (methotrexate only) was started at 10 days of life [17]. His cardiac functions stabilized by 4 weeks on this regimen and normalized within 3 months. His CK never rose above 900, and down trended after the 3rd infusion and has been normal since then. He had mild motor delay on exam at 6 months old.

LOPD: Thirty-one newborns were identified to have LOPD (Table 2). Twelve of them (39%) were homozygous for the most common splice site mutation (c.-32-13T>G). A total of 16 patients were compound heterozygous, and 11 of them carried the splice site mutation (c.-32-13T>G). Three cases were homozygous for c.2238G>C mutation.

Table 2. Biochemical Parameters of Screened Patients confirmed IOPD and LOPD.

	Newborn Screening				Confirmatory Testing			
	GAA#1 (>2.10 micromole/L/h)	GAA#2 (>2.10 micromole/L/h)	Genotype	GAA	AST and ALT (Normal Range for Lab, U/L)	CK (Normal Range for Lab, U/L)	BNP (Normal Range for Lab, pg/mL)	Hex4 (Normal Range for Lab, (mmol/mol Creatinine)
IOPD Patients								
1	0.65	0.22	c.759delC/c.1551+1G>C	4.4 (>6.7)	AST: 103 (30–100)	923 (60–305)	1080 (0.0–100.0)	24.7
2	0.65	0.34	c.525delT/c.1694_1697delTCTC	0.4 (<3.88)	AST: 189 (22–71); ALT: 91 (7–50)	846 (60–305)	1232.5 (0.0–100.0)	30.2 (<20)
LOPD Patients								
1	1.86	0.78	c.-32-13T>G/c.-32-13T>G	3.1 (>6.7)	AST: 72 (24–72); ALT: 41 (17–63)	530 (28–300)	ND	4.9 (≤20)
2	1.86	1.44	c.2238G>C/c.1552-3C>G	5.3 (>6.7)	AST: 21 (24–72); ALT: 23 (17–63)	72 (28–300)	ND	4.2 (≤20)
3	1.73	1.08	c.-32-13T>G/c.-32-13T>G	3.6 (>6.7)	AST: 56 (24–72); ALT: 49 (17–63)	125 (28–300)	ND	4.5 (≤20)
4	1.24	0.67	c.-32-13T>G/c.-32-13T>G	2.4 (>6.7)	AST: 66 (24–72); ALT: 52 (17–63)	361 (28–300)	ND	3.9 (≤20)
5	1.82	0.94	c.-32-13T>G/c.-32-13T>G	6.9 (>6.7)	AST: 52 (24–72); ALT: 46 (17–63)	333 (28–300)	ND	4.4 (≤20)
6	1.32	0.88	c.-32-13T>G/c.-32-13T>G	0.7 (>6.7)	AST: 61 (24–72); ALT: 39 (17–63)	176 (28–300)	ND	5.9 (≤20)
7	1.05	0.19	c.-32-13 T>C/c.546 G>A	2.2 (>3.88)		283 (30–135)	ND	2.6 (<3.0)
8	0.83	0.85	c.-32-13 T>G/c. 1655 T>C	0.93 (>3.88)		365 (60–305)	ND	5.23 (<−8.9)
9	0.97	0.33	c.2238G>C/c.2281delGinsAT/c.2065G>A	0.93 (>3.88)	AST: 72 (22–71); ALT: 43 (7–50)	145 (60–305)	13.0 (0.0–100.0)	5.3 (≤20)
10	0.84	0.49	c.1438-1G>C/c.-32-13T>G	2.1 (>3.88)	AST: 168 (20–64); ALT: 104 (12–42)	617 (60–305)	-	6.6 (≤20)
11	-	0.63	c.-32-13T>G/c.2560C>T	3.00 (>3.88)	AST: 125 (22–71); ALT: 50 (7–50)	421 (60–305)	-	-
12	-	-	c.-32-13T>G /c.2238G>C	2.60 (>3.88)	AST: 112 (22–71); ALT: 50 (7–50)	464 (60–305)	51.6 (0.0–100.0)	6.1 (≤20)
13	0.56	0.62	c.-32-13T>G/c.-32-13T>G	3.80 (>3.88)	AST: 78 (20–64); ALT: 42 (12–42)	122 (60–305)	71.8 (0.0–100.0)	4.8 (≤20)

Table 2. *Cont.*

	Newborn Screening				Confirmatory Testing			
	GAA#1 (>2.10 micromole/L/h)	GAA#2 (>2.10 micromole/L/h)	Genotype	GAA	AST and ALT (Normal Range for Lab, U/L)	CK (Normal Range for Lab, U/L)	BNP (Normal Range for Lab, pg/mL)	Hex4 (Normal Range for Lab, (mmol/mol Creatinine)
14	1.68	0.79	c.156_157delTC/c.-32-13T>G	2.40 (>3.88)	AST: 166 (20–64); ALT: 73 (12–42)	467 (60–305)	-	7.0 (≤20)
15	1.11	0.5	c.2238G>C/c.-32-13T>G/*c.2065G>A*	1.60 (>3.88)	AST: 49 (20–64); ALT: 27 (12–42)	68 (60–305)	-	2.8 (≤20)
16	1.06	0.26	c.-32-13T>G/c.-32-13T>G	6.4 (>3.88)	AST: 73 (22–71); ALT: 39 (7–50)	140 (60–305)	99.8 (0.0–100.0)	8.5 (≤20)
17	1.04	0.77	c.-32-13T>G/c.-32-13T>G	1.80 (>3.88)	AST: 150 (22–71); ALT: 42 (7–50)	275 (60–305)	44.5 (0.0–100.0)	5.2 (≤20)
18	0.7	0.34	c.-32-13T>G/c.2560C>T	1.40 (>3.88)	AST: 152 (22–71); ALT: 55 (7–50)	542 (60–305)	16.5 (0.0–100.0)	7.6 (≤20)
19	1.07	0.38	c.456_458insGA/c.-32-13T>G	1.10 (>3.88)	AST: 158 (22–71); ALT: 39 (7–50)	510 (60–305)	<10.0 (0.0–100.0)	8.1 (≤20)
20	0.88	0.77	c.-32-13T>G/c.-32-13T>G	3.50 (>3.88)	AST: 63 (20–64); ALT: 34 (12–42)	204 (60–305)	65.4 (0.0–100.0)	-
21	1.77	0.84	c.-32-13T>G/c.-32-13T>G	-	AST: 71 (22–71); ALT: 60 (7–50)	155 (60–305)	-	12.6 (≤20)
22	1.07	0.92	c.-32-13T>G/c.-32-13T>G	1.60 (3.88)	AST: 92 (22–71); ALT: 46 (7–50)	265 (60–305)	81.5 (0.0–100.0)	5.9 (≤20)
23	1.12	0.62	c.2238G>C; *c.2065G>A*/c.2238G>C; *c.2065G>A*	2.5 (3.88)	AST: 73 (22–71); ALT: 28 (7–50)	48 (60–305)	102.3 (0.0–100.0)	-
24		0.43	c.1856G>A/c.2238G>C			197 (60–305)		
25		0.95	c.2238G>C/c.2238G>C					
26		1.01	c.2238G>C/c.2238G>C					
27	1.1	0.42	c.1441T>C/c.2238G>C/*c.2065G>A*	2.40 (>3.88)	AST: 32 (22–71); ALT: 29 (7–50)	108 (60–305)	35.7 (0.0–100.0)	3.3 (≤20)
28		1.06	c.-32-13T>G/c.2238G>C/*c.2065G>A*					
29		0.48	c.-32-13T>G/c.1547G>A					
30		1.07	c.32-13T>G/c.32-13T>G					
31		0.94	c.32-13T>G/c.2238G>C /*c.2065G>A*					

Pathogenic or likely pathogenic variants are bolded; Pseudodeficiency mutations are made red and italicized. Transcript number: NM_000152.3; Genome build: GRCh37.

Table 3. Biochemical Parameters of Screened Patients defined as Suspected LOPD.

Suspected LOPD Patients	Newborn Screening				Confirmatory Testing			
	GAA#1 (>2.10 micromole/L/h)	GAA#2 (>2.10 micromole/L/h)	Genotype	GAA	AST and ALT (normal range for lab, U/L)	CK (normal range for lab, U/L)	BNP (normal range for lab, pg/mL)	Hex4 (normal range for lab, (mmol/mol creatinine)
1	0.91	0.2	c.-32-13T>G/c.692+3G>C	2.1	AST: 30 (24–72); ALT: 30 (17–63)	150	ND	3.3
2	2.06	1.99	c.-32-13T>G/c.1594G>A	3.8	AST: 41 (24–72); ALT: 30 (17–63)	134	ND	3.0
3	1.48	0.67	c.-32-13T>G/c.546G>A	1.6	AST: 42 (24–72); ALT: 37 (17–63)	ND	ND	6.0
4	1.85	0.98	c.-32-13T>G/c.266G>A;c.1377C>G	1.4	AST: 39 (24–72); ALT: 32 (17–63)	293	ND	8.3
5	1.04	0.53	c.-32-13T>G/c.2003 A>G	0.6 (>3.88)		107 (26-192)		2.4 (<20)
6	0.61	0.41	c.-32-13T>G/c.1721T>C	0.8	AST: 96 (24–72); ALT: 84 (17–63)	377	ND	4.9
7	0.52	0.45	c.-32-13T>G/c.1291_1299del9	1.9	AST: 83 (20–70); ALT: 48 (17–63)	486	ND	9.4
8	0.56	0.63	c.-32-13T>G/1655T>C	2.5	AST: 156 (24–72); ALT: 103 (17–63)	668	ND	6.0
9	ND	1.34	c.-32-13T>G/c.533G>A	3.7	AST: 25 (24–72); ALT: 19 (17–63)	86	ND	2.5
10	1.3	2.01	c.-32-13T>G/ c.862G>A;c.271G>A	2.0	AST: 32 (24–72); ALT: 31 (17–63)	406	ND	3.4
11	1.58	0.64	c.-32-13T>G/ c.841C>T	1.4	AST: 23 (24–72); ALT: 42 (17–63)	136	ND	6.4
12	0.58 umol/L/h	1.84 umol/L/h	c.1A>G/c.1345C>T	1.90 (>3.88)	AST: 44 (15–41 U/L) ALT: 26 (12–42)	241 (15-200 U/L)	-	8.4 (≤20 mmol/mol creatinine)
13	1.02 umol/L/h	1.22 umol/L/h	c.2560C>T/c.1888+5G>T;c.2065G>A	3.80 (3.88)	AST: 112 (20–64); ALT: 29 (12–42)	69 (60–305)	407.3 (0.0–100.0)	4.6 (≤20 mmol/mol creatinine)
14	<0.19 umol/L/h	0.12 umol/L/h	c.2236T>C/c.700A>G/C	0.90 (>3.88)	AST: 52 (22–71); ALT: 36 (7–50)	142 (60–305)	44.2 (0.0–100.0)	15.7 (≤20)
15	1.94 umol/L/h	2.06 umol/L/h	c.1478C>T/1194+3G>C	6.05 (>3.88)	-	86 (60–305)	-	<4.4 (≤20)
16	1.97 umol/L/h	1.16 umol/L/h	c.784G>A/c.859-19G>A/c.1392G>C	4.00 (>3.88)	AST: 92 (20–64); ALT: 42 (12–42)	201 (60–305)	55.3 (0.0–100.0)	6.1 (≤20)

Table 3. *Cont.*

Suspected LOPD Patients	Newborn Screening				Confirmatory Testing			
	GAA#1 (>2.10 micromole/L/h)	GAA#2 (>2.10 micromole/L/h)	Genotype	GAA	AST and ALT (normal range for lab, U/L)	CK (normal range for lab, U/L)	BNP (normal range for lab, pg/mL)	Hex4 (normal range for lab, (mmol/mol creatinine)
17	0.98 umol/L/h	0.58 umol/L/h	**c.2105G>T**/c.1124G>A	1.20 {>3.88)	AST: 64 (22–71); ALT: 20 (7–50)	85 (60–305)	-	3.0 (≤20)
18	1.59 umol/L/h	0.95 umol/L/h	**c.-32-13T>G**/c.692+3G>C	2.70 (>3.88}	AST:73 (22–71); ALT:19 (7–50)	65 (60–305)	52.7 (0.0–100.0)	3.3 (≤20)
19	1.19	1.27	**c.1655T>C**/c.1888+5G>T/*c.2065G>A*	0.43 (1.29-25.7)	AST: 60 (22–71); ALT: 25 (7–50)	110 (60–305)	19.3 (0.0–100.0)	-
20	-	1.27	**c.-32-13T>G**/c.705G>A/*c.(1726G>A;c.2065G>A)*	1.80 (3.88)	AST 80 (22–71); ALT: 41 (7–50)	86 (60–305)	30.0 (0.0–100.0)	3.3 (≤20)
21	1.64	1.1	**c.-32-13T>G**/c.650C>T	4.10 (>3.88)	AST: 65 (22–71); ALT: 35 (7–50)	68 (60–305)	33.8 (0.0–100.0)	5.7 (≤20)
22	1.02	0.85	**c.1655T>C**/c.664G>A	2.01 (>3.88)	AST:69 (20–64); ALT: 41 (12-42)	94 (60–305)	57.4 (0.0–100.0)	6.2 (≤20)
23	-	0.42	**c.-32-13T>G**/c.1103G>A	-	AST: 126 (20–64); ALT:85 (12–42)	407 {60–305)	-	7.2 (≤20)
24	-	1.08	**c.258dupC**/c.1909C>A/*c.(1726G>A;c.2065G>A)*	-	AST: 64 (22–71); ALT: 38 (7–50)	102 (60–305)	13.0 (0.0–100.0)	3.4 (≤20)
25	1.98	1.17	**c.1552-3C>G**/c.1378G>A	-	AST: 91 (22–71); ALT: 46 (7–50)	123 (60–305)	29.1 (0.0–100.0)	3.7 (≤20)
26	-	1.49	**c.1655T>C**/c.266G>A	2.10 (>3.88)	AST: 56 (22–71); ALT: 36 (7–50)	88 3 60–305)	-	3.8 (≤20)
27		0.42	**c.-32-13T>G**/c.2467A>T					
28		1.63	**c.1504A>G**/c.2467A>T					
29		1.36	**c.2560C>T**/c.726G>A					
30		0.93	**c.2238G>C**/c.2467A>T					

Transcript number: NM_000152.3; Genome build: GRCh37. Pathogenic or likely pathogenic variants are bolded; VUSs are normal; Pseudodeficiency mutations are made red and italicized; benign or likely benign are bolded and in blue. Variants were checked in the Pompe Disease Mutation Database, available at http://www.pompevariantdatabase.nl or https://www.ncbi.nlm.nih.gov/clinvar.

All patients had low confirmatory GAA enzyme activity except two patients who had levels within normal range (patient 5, and patient 16 in Table 2). Both of these patients were homozygous for the c.-32-13T>G mutation. CK levels ranged from 48 to 617 U/L, with 11 individuals demonstrating mildly elevated CK levels. Urine Hex4 levels were within normal range in all cases. All but 2 cases had ECGs at the initial visit. ECHO was done at the time of initial visit in 24 patients. Two cases had their first ECHO at 3 months and 6 months, respectively. One case showed mild ventricular septal hypertrophy and mild hypertrophy of the left ventricle at the time of confirmatory test but follow up ECG and ECHO were normal at 3 months of age. Otherwise, ECG and ECHO at the time of diagnosis were normal or demonstrated nonspecific findings. Chest radiograms were done only by one referral site, and they were normal.

Suspected LOPD: Thirty patients were compound heterozygous for one pathogenic variant and one variant of unknown significance (VUS) or two VUS and were defined as suspected LOPD (Table 3). A total of 16 of 30 carried the most common splice site (c.-32-13T>G) mutation. The confirmatory GAA enzyme activity was lower than normal in most cases; three individuals had levels within the low-end of the normal range (Table 3). CK levels ranged from 65 to 668 U/L. Six cases had slightly elevated CK levels at the time of initial visit. Urine Hex4 levels were normal in all suspected LOPD patients. ECGs were normal or showed non-specific findings. A total of 20 cases had ECHO at the time of diagnosis, and they were normal demonstrated non-specific findings such as PFO or secundum ASD.

Pseudodeficiency: Fifteen infants had one or two pseudodeficiency alles, and they were reported as pseudodeficiency carriers. Although a referral to a metabolic center was not made, PCPs were given an option to consult with a metabolic physician about these cases. A total of 8 of 15 newborns with pseudodeficiency were referred to metabolic centers to enhance families' understanding of the meaning of pseudodeficiency alleles.

Carriers: Thirty-five newborns had low enzyme activity and one mutation. Further work up, including deletion and duplication testing, confirmed that they were carriers for Pompe disease. Parents of these infants received genetic counseling in each case.

False positive: There were two newborns who had GAA enzyme activities 1.78 and 1.61 (>2.0 micromole/L/h) on the second NBS, respectively, but they had negative *GAA* gene sequencing. They were seen at a metabolism referral center, and their confirmatory enzyme levels were 4,6 (>3.88) and 5.6 (>3.88), respectively. Both were reported as false positive.

Questions raised by Screening Results

Biomarker data collected through NBS for Pompe disease raised several important questions.

Do NBS GAA and confirmatory GAA levels make it possible to distinguish different types of Pompe disease, pseudodeficiencies, and carriers?

The median level of GAA in LOPD patients was lower than those of suspected LOPD, carriers, and pseudodeficiency cases. Patients with LOPD had significantly lower levels of GAA enzyme activity compared to cases who are carriers or have pseudodeficiency alleles in all NBS and confirmatory tests ($p < 0.003$, 0 and 0.007). There was no statistically significant difference in GAA levels between LOPD and suspected LOPD cases except the second NBS ($p < 0.015$). Patients with suspected LOPD also had statistically lower GAA levels in all NBS and confirmatory tests compared to those of carriers/pseudodeficiency ($p < 0.029$, 0.002, 0.001) (Figure 2).

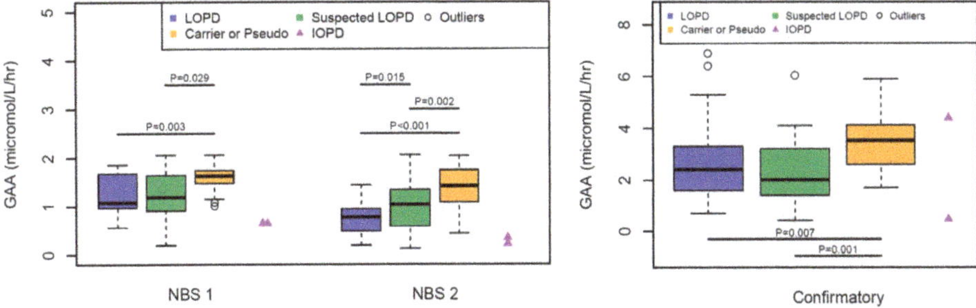

Figure 2. Comparing NBS and Confirmatory GAA enzyme activities of LOPD to Carriers/Pseudodeficiency or to Suspected LOPD patients, and suspected LOPD to carriers/pseudodeficiency) using Wilcoxon rank sum test. **NBS GAA#1** (Median: Min, Max): LOPD (*n*:22) (1.08:0.56,1.86); Suspected LOPD (*n*:21) (1.19: 0.19, 2.06); Carriers or Pseudo (*n*:30) (1.62: 1.01, 2.06), **NBS GAA#2** (Median: Min, Max): LOPD (*n*:30) (0.77:0.19, 1.44); Suspected LOPD (*n*:30) (1.03:0.12, 2.06); Carriers or Pseudo (*n*:43) (1.42:0.43, 2.04), **Confirmatory GAA** (Median, Min, Max): LOPD (*n*:23) (2.4:0.7, 6.9); Suspected LOPD (*n*:23) (2: 0.43, 6.05); Carriers or Pseudo (*n*:28) (3.5:1.7, 12.4).

Answer: In our cohort, the median value of GAA enzyme levels were statistically different, and it was possible to differentiate LOPD and suspected LOPD cases from pseudodeficiency and carriers. Two IOPD cases had much lower NBS GAA enzyme values compared to others. Although there were statistically significant differences, it is not always possible to utilize GAA levels to identify the Pompe disease status in individual newborns because of overlapping GAA values.

Do initial AST, ALT, CK, and Hex4 levels make it possible to distinguish different types of Pompe disease?

The median values of AST (72.5 U/L) and ALT (42.5 U/L) were higher in LOPD patients compared to those of suspected LOPD (AST: 64 U/L, ALT: 36 U/L) and carriers and pseudodeficiency (AST: 63.5 U/L, ALT: 32 U/L). The statistical difference was only noted in ALT values between LOPD and suspected LOPD ($p < 0.048$), and LOPD and carriers/pseudodeficiency ($p < 0.002$) (Figure 3).

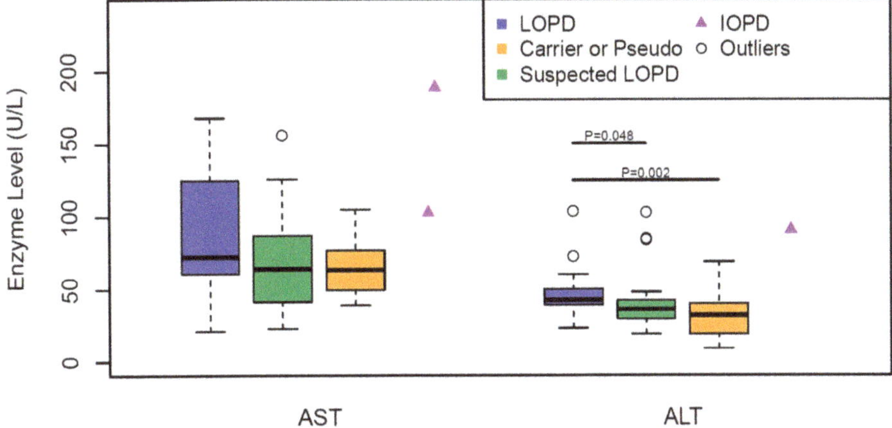

Figure 3. Comparing AST and ALT values of LOPD to Carriers/Pseudodeficiency or to Suspected LOPD patients, and suspected LOPD to carriers/Pseudodeficiency using Wilcoxon rank sum test. **AST** (U/L) (Median:Min, Max): LOPD (n:22) (72.5:21.1,168); Suspected LOPD (*n*:24) (64:23, 156); Carriers or Pseudo (*n*:20) (63.5:39, 105), **ALT**(U/L): LOPD (*n*:22) (42.5:23.1, 104); Suspected LOPD (*n*:24) (36:19, 103); Carriers or Pseudo (*n*:22) (32:9, 69).

The median value of CK (265 U/L) was significantly higher in LOPD patients compared to the median values of the suspected LOPD (123 U/L) or carriers/pseudodeficiency (107) ($p < 0.034$ and 0, respectively) (Figure 4).

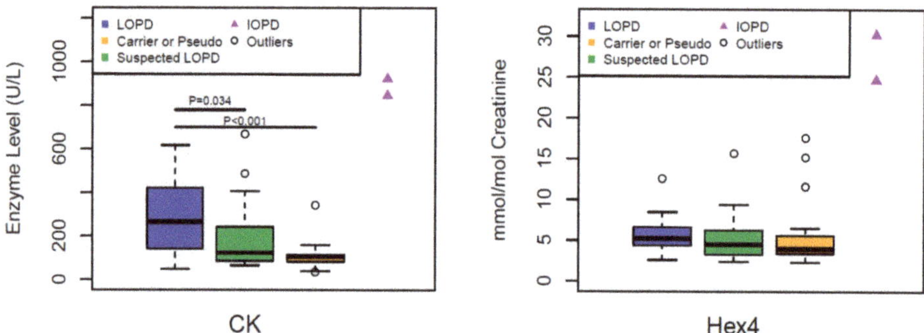

Figure 4. Comparing CK and Hex4 levels of LOPD to Carriers/Pseudodeficiency or to Suspected LOPD patients, and suspected LOPD to Carriers/Pseudodeficiency using Wilcoxon rank sum test. **CK** (U/L) (Median:Min, Max): LOPD (*n*:25) (265:48, 617); Suspected LOPD (*n*:25) (123:65, 668); Carriers or Pseudo (*n*:27) (107:35, 344), **Hex4** (mmol/mol creatinine): LOPD (*n*:21) (5.23:2.6, 12.6); Suspected LOPD (*n*:24) (4.5:2.4, 15.7); Carriers or Pseudo (*n*:26) (3.95:2.3, 17.6).

Although the median value of Hex 4 (5.23 mmol/mol creatinine) was higher in LOPD, it was not statistically different than those of suspected LOPD (4.5 mmol/mol creatinine) or carrier/pseudodeficiency (3.95 mmol/mol creatinine) (Figure 4).

None of these biomarkers were significantly different between suspected LOPD and carriers/pseudodeficiency (Figures 3 and 4).

Answer: In our cohort, the median values of CK and ALT (but not Hex4 or AST) levels were significantly higher in LOPD compared to other forms of Pompe disease detected by NBS. Two IOPD cases had much higher CK, AST, ALT and Hex4 levels compare to others. It is important to note that the analytes alone may not be used to make phenotyping or clinical decision in all cases.

Do cases homozygous for the common splice site mutation (c.-32-13T>G) have different GAA enzyme or CK levels compared to other LOPD patients?

The median values of GAA enzyme activity (NBS#2: 0.81 µmol/L/h, Confirmatory GAA: 3.3) were higher, and the median value of CK level (204 U/L) was lower in the patients homozygous for c.-32-13T>G mutation than those of other LOPD patients (NBS#2: 0.56 µmol/L/h, confirmatory GAA:2.2, CK:324 U/L), but we could not find any statistically significant difference (Figure 5).

Answer: No, GAA enzyme or CK levels do not help to differentiate cases homozygous for common splice site mutation from LOPD cases caused by other pathogenic mutations.

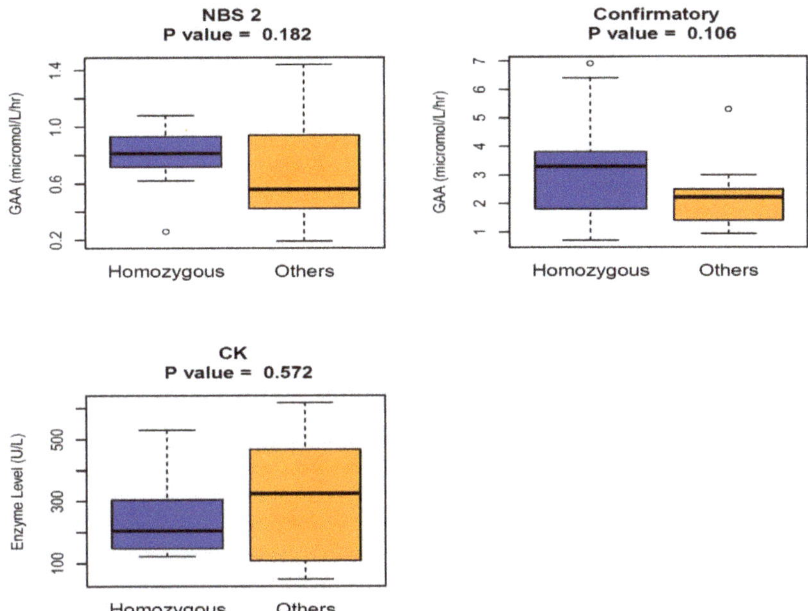

Figure 5. Comparison of NBS #2, Confirmatory GAA enzyme activities, and CK between c.-32-13T>G mutation homozygous patients and others in LOPD group, using Wilcoxon rank sum test. **NBS GAA#2** (Median:Min, Max): LOPD (homozygous for c.-32-13T>G) (*n*:12) (0.81:0.26, 1.08); LOPD-other pathogenic mutations (*n*:18) (0.56:0.19, 1.44); **Confirmatory GAA**: LOPD (homozygous for c.-32-13T>G) (*n*:10) (3.3:0.7, 6.9); LOPD-other pathogenic mutations (*n*:13) (2.2:0.93, 5.3); CK (U/L): LOPD (homozygous for c.-32-13T>G) (*n*:11) (204:122, 530); LOPD-other pathogenic mutations (*n*:14) (324:48, 617).

4. Discussion

Newborn screening detects all subtypes of Pompe disease. The reported incidence of the disease has increased since initiation of NBS, with values ranging from 1:8684 to 1:23,596 [23,26,27]. Increased incidence is largely due to improved detection of LOPD and suspected LOPD cases through NBS. In Pennsylvania, we found a similar increase with a combined incidence of IOPD and LOPD of 1:16,095 and a combined incidence of IOPD, LOPD, and suspected LOPD of 1:8,431. The incidence of IOPD by itself was 1:265,570 in Pennsylvania.

NBS for Pompe is done either by measuring only GAA enzyme activity or both GAA enzyme activity and full gene sequencing. In Pennsylvania, we chose to perform both GAA enzyme activity and gene sequencing in the NBS to increase certainty of diagnosis and provide a more detailed definition of positive screening at the initial patient visit [28]. Performing genetic testing as part of NBS has multiple advantages. The genetic test identifies those with pseudodeficiency alleles that are common in Pompe. Such patients do not need referral to a metabolic center for additional workup. Having a molecular diagnosis at the time of first visit informs the discussion of the different forms of the disease. It decreases anxiety by offering families a final diagnosis and a follow up plan based on the NBS test result. The availability of genotype at the time of reporting of abnormal NBS can also help in predicting CRIM status in IOPD patients and facilitates an informed and swift decision about the need for immunomodulation. In the absence of the results of genotype testing, families are subjected to uncertainty of diagnosis during the minimum two-week period required to order and receive the results of the genetic test. Including genetic testing with NBS for Pompe also circumvents the risk that insurance companies might deny coverage for it and guarantees that patients do not bear any additional financial burden.

NBS detects both IOPD and LOPD. Detecting IOPD cases in the first weeks of life is essential for initiation of therapy to yield optimal outcomes [21,29]. In our cohort, 2 patients with IOPD were detected (the incidence of IOPD by itself was 1:265,570 in Pennsylvania) and treated with ERT. The first patient was in the NICU of a local hospital and evaluated for feeding difficulty and cardiomyopathy. The physicians started a diagnostic work up to rule out potential etiologies and waited for the final result of Pompe NBS, which showed that he had IOPD at 19 days of age. This case showed us that there is a risk of delaying diagnosis of IOPD in the current Pompe algorithm in Pennsylvania. A second specimen is necessary to complete the sequencing test due to depletion of blood on the initial specimen on which all other newborn screening tests are performed. After detecting the first case of IOPD, we updated the NBS algorithm in Pennsylvania for Pompe to minimize delays in diagnosis of IOPD. The updated algorithm requires department of health (DOH) nurses to ask pediatricians if newborn has cardiomyopathy or symptoms of Pompe disease such as hypotonia and feeding issues when the first abnormal test is reported.

We identified 31 patients with LOPD. In addition to molecular testing results, biochemical findings such as GAA enzyme activity and CK level may help to differentiate LOPD cases from suspected LOPD or carriers/pseudodeficiency at the initial visit. In general, in LOPD cases as compared to suspected LOPD and carriers/pseudodeficiency carriers, GAA enzyme activities were lower, and CK levels were higher. Hex 4 levels were elevated and helpful biomarker for IOPD cases, but it was normal in LOPD cases at the time of diagnosis.

Since LOPD patients may present with symptoms at any age from infancy to adulthood, some may question the benefits and consequences of detection of LOPD patients via NBS. Early detection may increase family anxiety [30]. However, genetic counseling and clear explanations about LOPD along with a detailed follow-up plan reduce family anxiety accompanying initial diagnosis. We know that an extended delay in diagnosis occurs in most LOPD patients, which adversely impacts outcomes. Outcomes of early treatment of LOPD patients detected by NBS in Taiwan have been promising (21). Several metabolic centers in Pennsylvania closely follow patients with LOPD detected by NBS to make judgments about when to start treatment based on very early signs and symptoms of disease (biomarkers and developmental tests) in each individual patient.

The leaky *GAA* splice site variant, c.-32-13T>G in intron 1 is found on at least one allele in 68–90% of Caucasian patients [31]. In our NBS cohort, 74% had at least one c.-32-13T>G variant. There was no statistically meaningful difference in GAA enzyme or CK levels at the time of diagnosis between LOPD patients homozygous for c.-32-13T>G and other pathogenic mutations. Patients homozygous for c.-32-13T>G usually develop mild LOPD. However, recent reports described patients with more classical LOPD and highlighted the risk for arrhythmias in these patients [32–34]. These reports emphasize the importance of close follow up and detailed evaluations in all Pompe patients regardless of genotype.

We also identified 30 patients with suspected LOPD, i.e., they harbor two VUS or one pathogenic allele and one VUS. In this group, average GAA enzyme activity was significantly lower than those of carriers and cases with pseudodeficiency alleles on both NBS and confirmatory tests. The values of biomarkers such as CK, Hex4, AST, and ALT did not differ from those seen in carriers or pseudodeficiency carriers. This group poses a significant challenge. Some of these cases might never develop symptoms of Pompe disease; however, given our inability to fully assess disease risk in this cohort, they require ongoing monitoring. We follow up these patients every 6 months.

Pennsylvania data differ from those of Illinois and Missouri. In Illinois, a total of 684,290 infants were screened between 3 November 2014 and 30 September 2019 [27]. A total of 395 newborns with positive NBS were referred to metabolic centers. A total of 234 of 395 (59%) were false positive (normal confirmatory enzyme activity), and 62 (26%) carried pseudodeficiency alleles; this compares to only 1.7% and 13% in Pennsylvania, respectively. In Missouri, 467,000 newborns were screened for Pompe disease, and 274 had a positive test based on decreased GAA activity between January 2013 and December 2018 [35]. A total of 97 of 274 (35%) were false positive based on normal GAA activity, and 53 of 274 (19%) carried pseudodeficiency alleles; this compares to only 1.7% and 13% in Pennsylvania, respectively. These differences may arise from the fact that (1) Pennsylvania measures GAA activity twice on two different samples; Illinois and Missouri only test once; (2) each state makes its own decision about the chosen cutoff value for GAA; and (3) each state has different percentages of ethnic groups in the population.

5. Conclusions

The Pennsylvania experience shows that the overall incidence of Pompe cases increased after initiation of NBS compared to previously reported incidence [3–7] due to increased numbers of LOPD cases detected by screening. Furthermore, in our experience, NBS can detect suspected LOPD cases, some of whom may never go on to develop symptoms. Genotyping as a second-tier test was essential to inform the final diagnosis. This timely molecular diagnosis facilitated clear results disclosure at the first clinical visit and reduced parental anxiety. Finally, false positive and pseudodeficiency cases occurred at much lower rates than previously reported in other states.

Close monitoring and data collection on all patients detected through NBS is essential to assess the long-term outcomes and success of NBS for Pompe disease. NBS registries should be created and funded to enable data collection. Increased data collection will make it possible to identify and understand the pathogenic vs benign nature of VUS in cases defined as suspected LOPD.

Author Contributions: C.F., R.C.A.-N., J.B., S.R.C., B.S.D., P.L.G., N.H., C.M., N.L., and D.O. were involved in evaluation of patients with positive newborn screening tests, genetic counseling, and in treatment of those patients with a definitive diagnosis. J.C.D. performed and analyzed the newborn screening tests. R.X. analyzed the data and performed the statistical analysis. C.F., analyzed the data and wrote the first draft of manuscript. R.C.A.-N., J.B., S.R.C., B.S.D., J.C.D., P.L.G., N.H., C.M., N.L., D.O, and R.X. edited the manuscript. All authors have read and agreed to the published version of the manuscript.

Funding: This research received no external funding.

Acknowledgments: The authors would like to acknowledge Jordan S Shover and Stacey Gustin from Pennsylvania Department of Health, Division of Newborn Screening, and Bethany Sgroi from Perkin Elmer, for their help in collection of newborn screening data.

Conflicts of Interest: C.F. has received consulting fees and/or honoraria from Sanofi Genzyme, Biomarin, Takeda, Ultragenxy, Horizon, Orphazyme. He has received research funding from Takeda, Moderna, RegenxBio, Mallinckrodt, Biomarin, Sanofi-Genzyme, Orphan tech., Alexion, Homology Medicines. The rest of authors has no conflict of interest.

References

1. Hirschhorn, R.; Reuser, A.J. Glycogen storage disease type II: Acid alpha-glucosidase (acid maltase) deficiency. In *The Metabolic and Molecular Bases of Inherited Disease*, 8th ed.; Scriver, C.R., Beaudet, A.L., Sly, W.S., Childs, B., Beaudet, A., Valle, D., Kinzler, K.W., Vogelstein, B., Eds.; McGraw-Hill: New York, NY, USA, 2001; pp. 3389–3420.
2. Hesselink, R.P.; Wagenmakers, A.J.; Drost, M.R.; Van der Vusse, G.J. Lysosomal dysfunction in muscle with special reference to glycogen storage disease type II. *Biochim. Biophys. Acta* **2003**, *1637*, 164–170. [CrossRef]
3. Ausems, M.G.; Verbiest, J.; Hermans, M.P.; Kroos, M.A.; Beemer, F.A.; Wokke, J.H.; Sandkuijl, L.A.; Reuser, A.J.; van der Ploeg, A.T. Frequency of glycogen storage disease type II in The Netherlands: Implications for diagnosis and genetic counselling. *Eur. J. Hum. Genet.* **1999**, *7*, 713–716. [CrossRef]
4. Martiniuk, F.; Chen, A.; Mack, A.; Arvanitopoulos, E.; Chen, Y.; Rom, W.N.; Codd, W.J.; Hanna, B.; Alcabes, P.; Raben, N.; et al. Carrier frequency for glycogen storage disease type II in New York and estimates of affected individuals born with the disease. *Am. J. Med. Genet.* **1998**, *79*, 69–72. [CrossRef]
5. Meikle, P.J.; Hopwood, J.J.; Clague, A.E.; Carey, W.F. Prevalence of lysosomal storage disorders. *JAMA* **1999**, *281*, 249–294. [CrossRef] [PubMed]
6. Pinto, R.; Caseiro, C.; Lemos, M.; Lopes, L.; Fontes, A.; Ribeiro, H.; Pinto, E.; Silva, E.; Rocha, S.; Marcão, A.; et al. Prevalence of lysosomal storage diseases in Portugal. *Eur. J. Hum. Genet.* **2004**, *12*, 87–92. [CrossRef] [PubMed]

7.   Poorthuis, B.J.H.M.; Wevers, R.A.; Kleijer, W.J.; Groener, J.E.M.; de Jong, J.G.N.; van Weely, S.; Niezen-Koning, K.E.; van Diggelen, O.P. The frequency of lysosomal storage diseases in the Netherlands. *Hum. Genet.* **1999**, *105*, 151–156. [CrossRef] [PubMed]

8.   Kishnani, P.S.; Howell, R.R. Pompe Disease in infants and children. *J. Pediatr.* **2004**, *144*, S35–S43. [CrossRef] [PubMed]

9.   Young, S.P.; Zhang, H.; Corozo, D.; Thurberg, B.L.; Bali, D.; Kishnani, P.; Millington, D. Long-term monitoring of patients with infantile-onset Pompe disease on enzyme replacement therapy using a urinary glucose tetrasaccharide biomarker. *Genet. Med.* **2009**, *11*, 536–541. [CrossRef]

10.   Chien, Y.H.; Goldstein, J.L.; Hwu, W.; Smith, P.B.; Lee, N.; Chiang, S.; Tolun, A.A.; Zhang, H.; Vaisnins, A.E.; Millington, D.S.; et al. Baseline Urinary Glucose Tetrasaccharide Concentrations in Patients with Infantile- and Late-Onset Pompe Disease Identified by Newborn Screening. *JIMD Rep.* **2015**, *19*, 67–73.

11.   Van der Ploeg, A.T.; Reuser, A.J. Pompe disease. *Lancet* **2008**, *372*, 1342–1353. [CrossRef]

12.   Kishnani, P.S.; Amartino, H.M.; Lindberg, C.; Miller, T.M.; Wilson, A.; Keutzer, J. Timing of diagnosis of patients with Pompe disease: Data from the Pompe registry. *Am. J. Med. Genet. A* **2013**, *161*, 2431–2443. [CrossRef] [PubMed]

13.   Kishnani, P.S.; Corzo, D.; Nicolino, M.; Byrne, B.; Mandel, H.; Hwu, W.L. Recombinant human acid (alpha)-glucosidase: Major clinical benefits in infantile-onset Pompe disease. *Neurology* **2007**, *68*, 99–109. [CrossRef] [PubMed]

14.   Kishnani, P.S.; Goldenberg, P.C.; DeArmey, S.L.; Heller, J.; Benjamin, D.; Young, S.; Bali, D.; Smith, S.A.; Li, J.S.; Mandel, H.; et al. Cross-reactive immunologic material status affects treatment outcomes in Pompe disease infants. *Mol. Genet. Metab.* **2010**, *99*, 26–33. [CrossRef] [PubMed]

15.   Mendelsohn, N.J.; Messinger, Y.H.; Rosenberg, A.S.; Kishnani, P.S. Elimination of antibodies to recombinant enzyme in Pompe's disease. *N. Engl. J. Med.* **2009**, *360*, 194–195. [CrossRef] [PubMed]

16.   Messinger, Y.H.; Mendelsohn, N.J.; Rhead, W.; Dimmock, D.; Hershkovitz, E.; Champion, M.; Jones, S.A.; Olson, R.; White, A.; Wells, C.; et al. Successful immune tolerance induction to enzyme replacement therapy in CRIM-negative infantile Pompe disease. *Genet. Med.* **2012**, *14*, 135–142. [CrossRef]

17.   Kazi, Z.B.; Desai, A.K.; Troxler, R.B.; Kronn, D.; Packman, S.; Sabbadini, M.; Rizzo, W.B.; Scherer, K.; Abdul-Rahman, O.; Tanpaiboon, P.; et al. An immune tolerance approach using transient low-dose methotrexate in the ERT-naive setting of patients treated with a therapeutic protein: Experience in infantile-onset Pompe disease. *Genet. Med.* **2019**, *21*, 887–895. [CrossRef]

18.   Chien, Y.H.; Hwu, W.; Lee, N.C. Pompe Disease: Early Diagnosis and Early Treatment Make a Difference. *Pediatr. Neonatol.* **2013**, *54*, 219–227. [CrossRef]

19.   Chien, Y.H.; Lee, N.C.; Chen, C.A.; Tsai, F.S.; Tsai, W.H.; Shieh, J.Y.; Huang, H.J.; Hsu, W.C.; Tsai, T.H.; Hwu, W.L. Long-term prognosis of patients with infantile-onset Pompe disease diagnosed by newborn screening and treated since birth. *J. Pediatr.* **2015**, *166*, 985–991. [CrossRef]

20.   Chien, Y.H.; Lee, N.C.; Thurberg, B.L.; Chiang, S.C.; Zhang, X.K.; Keutzer, J.; Huang, A.C.; Wu, M.H.; Huang, P.H.; Tsai, F.J.; et al. Pompe Disease in Infants: Improving the Prognosis by Newborn Screening and Early Treatment. *Pediatrics* **2009**, *124*, e1116. [CrossRef]

21.   Chien, Y.H.; Lee, N.C.; Huang, H.S.; Thurberg, T.; Tsai, F.J.; Hwu, W.L. Later-Onset Pompe Disease: Early Detection and Early Treatment Initiation Enabled by Newborn Screening. *J. Pediatr.* **2011**, *158*, 1023–1027. [CrossRef]

22.   Chien, Y.H.; Hwu, W.L.; Lee, N.C. Newborn screening: Taiwanese experience. *Ann. Transl. Med.* **2019**, *7*, 281. [CrossRef] [PubMed]

23.   Bodamer, O.A.; Scott, C.R.; Giugliani, R. Newborn screening for Pompe disease. *Pediatrics* **2017**, *140*, S4–S13. [CrossRef] [PubMed]

24.   Kemper, A.R.; Browning, M. Evidence Review: Pompe Disease. Available online: https: //www.hrsa.gov/sites/default/files/hrsa/advisory-committees/heritable-disorders/rusp/previous-nominations/pompe-evidence-review-report-2008.pdf (accessed on 12 November 2020).

25.   Tortorelli, S.; Turgeon, C.T.; Gavrilov, D.K.; Oglesbee, D.; Raymond, K.M.; Rinaldo, P.; Matern, D. Simultaneous Testing for 6 Lysosomal Storage Disorders and X-Adrenoleukodystrophy in Dried Blood Spots by Tandem Mass Spectrometry. *Clin. Chem.* **2016**, *62*, 1248–1254. [CrossRef] [PubMed]

26. Mechtler, T.P.; Stary, S.; Metz, T.F.; de Jesús, V.R.; Greber-Platzer, S.; Pollak, A.; Herkner, K.R.; Streubel, B.; Kasper, D.C. Neonatal screening for lysosomal storage disorders: Feasibility and incidence from a nationwide study in Austria. *Lancet* **2012**, *379*, 335–341. [CrossRef]

27. Burton, B.K.; Charrow, J.; Hoganson, G.E.; Fleischer, J.; Grange, D.K.; Braddock, S.R.; Hitchins, L.; Hickey, R.; Christensen, K.M.; Groepper, D.; et al. Newborn Screening for Pompe Disease in Illinois: Experience with 684,290 Infants. *Int. J. Neonatal Screen.* **2020**, *6*, 4. [CrossRef]

28. Smith, L.D.; Bainbridge, M.N.; Parad, R.B.; Bhattacharjee, A. Second-Tier Molecular Testing in New-born Screening for Pompe Disease: Landscape and Challenges. *Int. J. Neonatal Screen.* **2020**, *6*, 32. [CrossRef]

29. Jang, C.F.; Jang, C.C.; Liao, H.C.; Huang, L.Y.; Chiang, C.C.; Ho, H.C.; Lai, C.H.; Chu, T.H.; Yang, T.F.; Hsu, T.R.; et al. Very Early Treatment for Infantile-Onset Pompe Disease Contributes to Better Outcomes. *J. Pediatr.* **2016**, *169*, 174–180.

30. Pruniski, B.; Lisi, E.; Ali, N. Newborn screening for Pompe disease: Impact on families. *J. Inherit. Metab. Dis.* **2018**, *41*, 1189–1203. [CrossRef]

31. Kroos, M.A.; Pomponio, R.J.; Hagemans, M.L.; Keulemans, J.L.M.; Phipps, M.; DeRiso, M.; Palmer, R.E.; Ausems, M.G.E.M.; van der Beek, N.A.M.E.; van Diggelen, O.P.; et al. Broad spectrum of Pompe disease in patients with the same c-32-13T->G haplotype. *Neurology* **2007**, *68*, 110–115. [CrossRef]

32. Rairkar, M.V.; Case, L.E.; Bailey, L.A.; Kazi, Z.B.; Desai, A.K.; Berrier, K.L.; Coats, J.; Gandy, R.; Quinones, R.; Kishnani, P. Insight into the phenotype of infants with Pompe disease identified by newborn screening with the common c.-32-13T>G "late-onset" GAA variant. *Mol. Genet. Metab.* **2017**, *122*, 99–107. [CrossRef]

33. Herbert, M.; Cope, H.; Li, J.S.; Kishnani, P. Severe Cardiac Involvement Is Rare in Patients with Late-Onset Pompe Disease and the Common c.-32-13T>G Variant: Implications for Newborn Screening. *J. Pediatr.* **2018**, *198*, 308–312. [CrossRef] [PubMed]

34. Musumeci, O.; Thieme, A.; Claeys, K.G.; Wenninger, S.; Kley, R.A.; Kuhn, M.; Lukacs, Z.; Deschauer, M.; Gaeta, M.; Toscano, A.; et al. Homozygosity for the common GAA gene splice site mutation c-32-13T>G in Pompe disease is associated with the classical adult phenotypical spectrum. *Neuromuscul. Disord.* **2015**, *25*, 719–724. [CrossRef] [PubMed]

35. Klug, T.; Swartz, L.B.; Washburn, J.; Brannen, C.; Kiesling, J. Lessons Learned from Pompe Disease Newborn Screening and Follow-up. *Int. J. Neonatal Screen.* **2020**, *6*, 11. [CrossRef] [PubMed]

Publisher's Note: MDPI stays neutral with regard to jurisdictional claims in published maps and institutional affiliations.

International Journal of
Neonatal Screening

Article

The Timely Needs for Infantile Onset Pompe Disease Newborn Screening—Practice in Taiwan

Shu-Chuan Chiang [1], Yin-Hsiu Chien [1,2,*], Kai-Ling Chang [1], Ni-Chung Lee [1,2] and Wuh-Liang Hwu [1,2]

[1] Department of Medical Genetics, National Taiwan University Hospital, Taipei 100, Taiwan
[2] Department of Pediatrics, National Taiwan University Hospital, Taipei 100, Taiwan
* Correspondence: chienyh@ntu.edu.tw; Tel.: +886-2-23123456 (ext. 71937); Fax: +886-2-23314518

Received: 9 February 2020; Accepted: 30 March 2020; Published: 1 April 2020

Abstract: Pompe disease Newborn screening (NBS) aims at diagnosing patients with infantile-onset Pompe disease (IOPD) early enough so a timely treatment can be instituted. Since 2015, the National Taiwan University NBS Center has changed the method for Pompe disease NBS from fluorometric assay to tandem mass assay. From 2016 to 2019, 14 newborns were reported as high-risk for Pompe disease at a median age of 9 days (range 6–13), and 18 were with a borderline risk at a median age of 13 days (9–28). None of the borderline risks were IOPD patients. Among the 14 at a high-risk of Pompe disease, four were found to have cardiomyopathy, and six were classified as potential late-onset Pompe disease. The four classic IOPD newborns, three of the four having at least one allele of the cross-reactive immunologic material (CRIM)-positive variant, started enzyme replacement therapy (ERT) at a median age of 9 days (8–14). Western Blot analysis and whole gene sequencing confirmed the CRIM-positive status in all cases. Here, we focus on the patient without the known CRIM-positive variant. Doing ERT before knowing the CRIM status created a dilemma in the decision and was discussed in detail. Our Pompe disease screening and diagnostic program successfully detected and treated patients with IOPD in time. However, the timely exclusion of a CRIM-negative status, which is rare in the Chinese population, is still a challenging task.

Keywords: infantile-onset Pompe disease; GAA sequencing; immune modulation therapy; enzyme replacement therapy; cross-reactive immunologic material

1. Introduction

Pompe disease, a genetic disorder caused by variants of the glucosidase alpha acid (*GAA*) gene, leads from acid alpha-glucosidase (GAA) deficiency. The phenotypes of Pompe disease vary widely, ranging from the most severe classic infantile-onset Pompe disease (IOPD) to the later-onset Pompe disease (LOPD). Currently, enzyme replacement therapy (ERT) with recombinant human GAA (rhGAA) is the only approved therapy. We have performed newborn screening (NBS) for Pompe disease [1] since 2005, and our results demonstrate that the early initiation of treatment improves the prognosis IOPD patients [2], thus confirming the value of newborn screening for Pompe disease.

However, there are still challenges when an infant receives a positive screening result. First, the phenotype cannot be predicted by GAA activity [3,4]. Mutation analysis of the *GAA* gene can predict the phenotype in a portion of patients [5]. But more precisely, newborns with classic IOPD should have presented with cardiomyopathy and muscle weakness at birth clinically [2,6]. Second, ERT with rhGAA may trigger an immune response with neutralizing antibodies, especially in patients negative for the cross-reactive immunologic material (CRIM) [7]. Prophylactic immunologic modulation therapy may overcome the problem, but the CRIM status needs to be defined before initiating ERT [8]. Nowadays, some GAA variants are associated with a known CRIM status [9].

The National Taiwan University Hospital (NTUH) Newborn Screening Center, established since 1985, is responsible for the screening of more than one-third of all newborns in Taiwan [10]. In 2005, we were the first to implement Pompe disease newborn screening [1]. Initially, dried blood spot (DBS) GAA activities were measured using fluorogenic (4-methylumbelliferone) substrates [4]. Since 2015, we have been using tandem mass assay (MS/MS) substrates in order to accommodate multiplexing ability [11]. The medical genetics department in the NTUH is also the referral center for Pompe disease detected by the NTUH. The hospital staff work closely with the screening center in order to make a timely management of IOPD. Here, we describe our practice in the past seven years.

2. Methods

NBS for inborn errors of metabolism in Taiwan was established in 1985, and, currently, the National Taiwan University Hospital (NTUH) holds one of the three screening centers in Taiwan. There are more than 300 birthing facilities that collect newborn dried bloodspots (DBS) and ship them promptly to the NBS labs. DBS sampling is usually performed 48–72 h after the birth of the babies, and shipping by priority mail typically takes less than two days. The NBS labs are requested to report high-risk results within 72 h after receiving the samples [10]. Although NBS is not mandatory in Taiwan, close to 100% of newborns acquired NBS. In 2008, Pompe disease newborn screening was added by the NTUH NBS Center and also by the other two screening centers [12], but written consent from the parent(s) is required. More than 95% of parents receiving service of NTUH NBS center provide consents for having Pompe disease newborn screening. The methods of Pompe disease NBS has been described previously [1,11,13]. Initially, GAA activity in DBS elute is measured using fluorogenic substrates, but the method was changed to the tandem mass spectrometry (MS/MS) at the end of 2015. From the first DBS, we set two cutoffs of the GAA activity measurements. Values exceeding the critical cutoff imply a high risk of having Pompe disease and that emergent confirmatory diagnostic testing is necessary. Values exceeding the borderline cutoff, mostly due to pseudodeficiency, will trigger a second-tier test, generating the value of % inhibition, before the final assignment [4]. If the second-tier test is positive, the baby will be suggested to have the confirmatory diagnostic testing. Since 2013, we employed the second-tier test to avoid requesting a second sample to prevent delay in the initiation of treatment. DBS DNA genotyping [14] may be applied to categorise the newborns. For confirmatory diagnostic testing, a whole blood sample was used for the measurement of GAA activity, genotyping, and CRIM test [10]. The CRIM test was performed by Western Blot analysis, using anti-GAA and anti-alpha-tubulin antibodies. For GAA protein detection, 15 μL of sonicated lymphocytes protein from patients was loaded in each lane (only 2 μL was used for normal control) and the X-ray film was exposed overnight. For alpha-tubulin (the control protein), 10 μL of sonicated protein was loaded in each lane and the X-ray film was exposed for 15 min. The Taiwanese common variant, p.D645E (p.Asp645Glu) [6,15], was rapidly screened by polymerase chain reaction-restriction fragment length polymorphism analysis (RFLP) using the *Bsa*HI restriction enzyme. Since p.D645E is a CRIM-positive variant [16], patients with this variant should be CRIM-positive. Enzyme replacement therapy (ERT) is scheduled for the next day after the heart involvement is confirmed, unless immunomodulation therapy to prevent anti-GAA antibodies production is planned. For babies without heart involvement at birth, a follow-up plan was initiated, including the development milestone, motor function, and biomarkers as described [10], and ERT was initiated until abnormalities appeared in the follow-up period.

3. Results

3.1. Performance of Screening and Diagnostic Testing

From 2016 to 2019, Pompe disease NBS was performed by the MS/MS method. The timeliness of the performance of other screening conditions (glucose-6-phosphate dehydrogenase deficiency, congenital hypothyroidism, galactosemia, congenital adrenal hyperplasia, and MS/MS acylcarnitine profile) during this period was similar to the previous three years (2013–2015); i.e., the compliance rates of reporting NBS results by 8 days of age were between 98.66% to 99.01%, compared to 98.85%–99.14% in the previous 3 years. With the MS/MS platform, 14 newborns were reported as high-risk for Pompe disease at a median age of 9 days (range 6–13). For newborns with a borderline risk for Pompe disease, 18 were reported at a median age of 13 days (9–28), but none of them were IOPD. During the previous 3 years (2013–2015), when Pompe disease NBS was performed using the 4MU platform, there were no Pompe disease high-risk newborns and 44 were reported as a borderline risk reported at a median age of 12 days (7–41), but none were IOPD.

Among the 14 newborns at a high risk of Pompe disease, four were found to have cardiomyopathy, as shown by electrocardiography, chest X-ray, echocardiography, and an elevation of serum creatine kinase (CK) and pro-brain natriuretic peptide (pro-BNP). Six [11,13] were found to have GAA deficiency and biallelic GAA variants but normal CK and no cardiomegaly and therefore were classified as potential LOPDs. The remaining four of the total 14 newborns were not affected. As for the 18 infants with the borderline risk, only 2 infants [11] were classified as potential LOPDs, while the rest were not affected. The four classic IOPD newborns were treated starting from a median of 9 days (8–14 days). *GAA* gene sequencing confirmed all pathogenic variants in these patients. Three of the four had at least one allele of the p.D645E variant. CRIM status (all CRIM positive (denoted as CRIM +)) was approved by Western Blot analysis in all four patients, including the one who did not have the p.D645E variation. The newborn who did not have the p.D645E variant did create a dilemma in the decision, and the history is described below.

3.2. Case Description

A 7-days-old female newborn was requested to visit our hospital due to an abnormal Pompe disease screening result [13]. She was born at full-term to a G1P1 mother with a birth bodyweight of 4380 gm. The parents denied poor feeding, poor activity, nor weak crying in this baby. NBS included Pompe disease, and other conditions were performed on her third day of life. On Day 4, our NBS laboratory received her sample. On Day 6, a high risk of Pompe disease was reported (GAA activity 0.18 uM/h (critical cutoff < 0.5); ratio 42.87 (critical cutoff acid β-glucosidase (ABG)/GAA ≥ 20) so an urgent visit to our hospital was arranged on Day 7. When we saw her, she had normal muscle power, normal reflex, no macroglossia, but her facial folds decrease slightly. Laboratory examination revealed an elevation of pro-BNP (8738 pg/mL), CK (722 U/L), and alanine aminotransferase (ALT) (112 U/L). A chest X-ray revealed mild cardiomegaly (Figure 1). Echocardiography revealed moderate left ventricle (LV) and right ventricle (RV) hypertrophy, with a LV mass index (LVMI, measured by 2-D method) of 115.7 g/m^2 (normal range < 65 g/m^2). A whole blood sampling at Day 7 revealed deficient lymphocyte GAA activity (1.33 nmol/g pro/h, normal mean 66.7) and thus confirmed the diagnosis of IOPD. However, she did not have the common CRIM-positive Taiwan Pompe disease p.D645E variant, tested using DNA extracted from the first DBS. Although her CRIM status was unknown, her parents refused prophylactic immune modulation therapy. Western Blot analysis using the white blood cells as the material soon revealed the 110 kDa precursor GAA band (Figure 2), suggesting a CRIM-positive status. She received her first dose of rhGAA (20 mg/kg) at the age of 8 days. Mutation analysis showed heterozygous c.2024_2026del (p.N675del) and c.2040+1G>T variants *in trans*, compatible with IOPD. There were no CRIM status predictions about these two variants [9].

The study was approved by the ethical committee of National Taiwan University Hospital, Taipei, Taiwan (201906053RINB, 1st approved date 2019/08/05). The clinical information was gathered from the hospital medical records retrospectively, and no individual's consent was required.

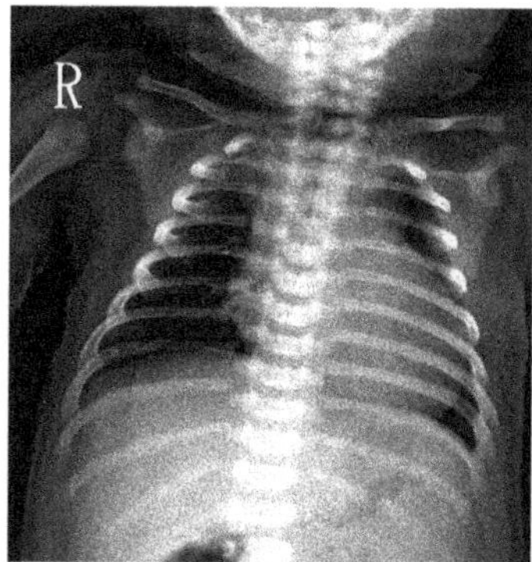

Figure 1. CXR at D7 in a newborn with a positive Pompe newborn screening result. Mild cardiomegaly was noted.

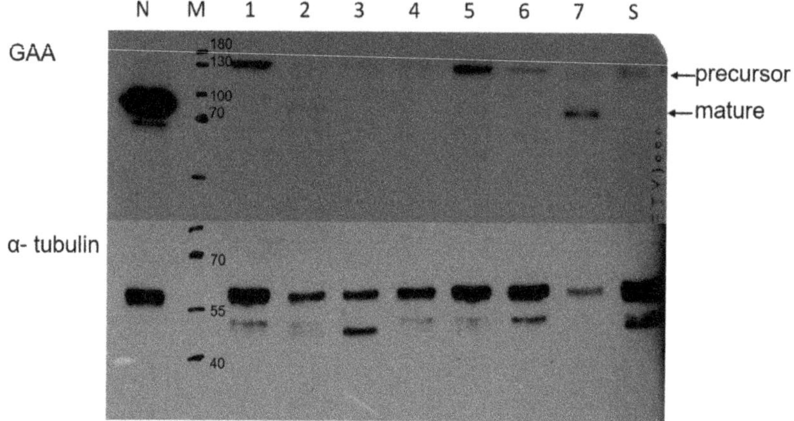

Figure 2. Blood glucosidase alpha acid (GAA) Western Blotting of this case (S), indicating a cross-reactive immunologic material (CRIM)-positive status. The lymphocytes were sonicated and then blotted with antibodies to detect human GAA and α-tubulin protein presence. M: marker; N: normal newborn; No. 1 and No. 3 were from infantile-onset Pompe disease (IOPD) patients with the CRIM(+) GAA variant; No. 2, No. 4–7 were from newborns with low GAA activity. N: normal control sample; M: marker. Precursor: 110 kDa GAA. Mature: 70/76 kDa GAA.

4. Discussion

Pompe newborn screening was included in Taiwan's newborn screening system as well as in the Recommended Uniform Screening Panel (RUSP) in the USA. Therefore, the timeliness requirements of newborn screening, i.e., the efficient collection, transportation, testing, and reporting of the results, also benefit Pompe newborn screening. In Taiwan, the recommended timelines for the high-risk babies is to report and communicate the results to the newborn's healthcare provider/parent(s) within 6–8 days of life, regardless of the location of newborns, which maybe 300 km away from the screening centers/treatment centers (Figure 3). The compliance rate for reporting by age 8 days in our center was over 99%. In the present case, we informed the result by D6 and made the diagnosis by D7, demonstrating the discrimination power of applying the critical cutoffs and the well-established newborn screening system in Taiwan.

Figure 3. The coverage map of the National Taiwan University Hospital Newborn Screening Center. The samples ship from the gray areas to our screening center, designated by the government, including the three far areas indicated by the lines. The distances from the birthplaces to the screening center were around 300 km.

The more challenging part of this case was the preparation of ERT, starting after the confirmation of cardiomegaly and muscle damage. In this case, the decision for ERT was tentatively made at D7 and the ERT was initiated at D8, after reconfirming the GAA deficiency and confirming the CRIM status. In such classic IOPD newborns waiting for the decision of prophylaxis immunomodulation, we routinely check the predicted CRIM status by screening the p.D645E, commonly seen in our Pompe patients [6], using RFLP so that we could have the result in half day. Since p.D645E is related to CRIM-positive status [16], patients with at least one p.D645E allele will be CRIM-positive, and doing prophylaxis immunomodulation on such cases may not achieve benefit-risk balance. On the other hand, we plan to apply prophylactic immune modulation therapy for IOPD infants if a CRIM-negative status is confirmed. Therefore, in this case, we performed the blood Western Blotting assay as described [17] to determine the CRIM status since the GAA sequencing result took more time, and the CRIM status for a novel variant may not be predicable, especially for the splicing mutation [9] presented in this case. Rapid sequencing, or screening for several common variants, may replace the blood CRIM status measurement in such a situation.

In conclusion, we demonstrate here the performance of the NBS system in Taiwan and the decision steps for positive Pompe NBS. Knowing the genotype/CRIM status was necessary for the ERT initiation but it made for very intensive work. Depending on the different geographic regions and various resources available, each team needs to prepare themselves with a standardized and comprehensive algorithm for the confirmation and treatment of classic IOPD patients. With the timeliness of screening and diagnosis, we were able to start the treatment as early as possible to achieve the best treatment outcome.

Author Contributions: Data curation, S.-C.C. and K.-L.C.; Formal analysis, Y.-H.C. and N.-C.L.; Methodology, S.-C.C., W.-L.H. All authors have read and agreed to the published version of the manuscript.

Funding: This research received no external funding.

Conflicts of Interest: The authors declare no conflict of interest.

References

1. Chien, Y.H.; Chiang, S.C.; Zhang, X.K.; Keutzer, J.; Lee, N.C.; Huang, A.C.; Chen, C.A.; Wu, M.H.; Huang, P.H.; Tsai, F.J.; et al. Early detection of Pompe disease by newborn screening is feasible: Results from the Taiwan screening program. *Pediatrics* **2008**, *122*, e39–e45. [CrossRef] [PubMed]
2. Chien, Y.H.; Lee, N.C.; Thurberg, B.L.; Chiang, S.C.; Zhang, X.K.; Keutzer, J.; Huang, A.C.; Wu, M.H.; Huang, P.H.; Tsai, F.J.; et al. Pompe disease in infants: Improving the prognosis by newborn screening and early treatment. *Pediatrics* **2009**, *124*, e1116–e1125. [CrossRef] [PubMed]
3. Chien, Y.H.; Lee, N.C.; Huang, H.J.; Thurberg, B.L.; Tsai, F.J.; Hwu, W.L. Later-onset Pompe disease: Early detection and early treatment initiation enabled by newborn screening. *J. Pediatr.* **2011**, *158*, 1023–1027.e1. [CrossRef] [PubMed]
4. Chiang, S.C.; Hwu, W.L.; Lee, N.C.; Hsu, L.W.; Chien, Y.H. Algorithm for Pompe disease newborn screening: Results from the Taiwan screening program. *Mol. Genet. Metab.* **2012**, *106*, 281–286. [CrossRef] [PubMed]
5. Reuser, A.J.J.; van der Ploeg, A.T.; Chien, Y.H.; Llerena, J., Jr.; Abbott, M.A.; Clemens, P.R.; Kimonis, V.E.; Leslie, N.; Maruti, S.S.; Sanson, B.J.; et al. GAA variants and phenotypes among 1,079 patients with Pompe disease: Data from the Pompe Registry. *Hum. Mutat.* **2019**, *40*, 2146–2164. [CrossRef] [PubMed]
6. Chien, Y.H.; Lee, N.C.; Chen, C.A.; Tsai, F.J.; Tsai, W.H.; Shieh, J.Y.; Huang, H.J.; Hsu, W.C.; Tsai, T.H.; Hwu, W.L. Long-term prognosis of patients with infantile-onset Pompe disease diagnosed by newborn screening and treated since birth. *J. Pediatr.* **2015**, *166*, 985–991. [CrossRef] [PubMed]
7. Banugaria, S.G.; Patel, T.T.; Mackey, J.; Das, S.; Amalfitano, A.; Rosenberg, A.S.; Charrow, J.; Chen, Y.T.; Kishnani, P.S. Persistence of high sustained antibodies to enzyme replacement therapy despite extensive immunomodulatory therapy in an infant with Pompe disease: Need for agents to target antibody-secreting plasma cells. *Mol. Genet. Metab.* **2012**, *105*, 677–680. [CrossRef] [PubMed]
8. Messinger, Y.H.; Mendelsohn, N.J.; Rhead, W.; Dimmock, D.; Hershkovitz, E.; Champion, M.; Jones, S.A.; Olson, R.; White, A.; Wells, C.; et al. Successful immune tolerance induction to enzyme replacement therapy in CRIM-negative infantile Pompe disease. *Genet. Med.* **2012**, *14*, 135–142. [CrossRef] [PubMed]
9. Bali, D.S.; Goldstein, J.L.; Banugaria, S.; Dai, J.; Mackey, J.; Rehder, C.; Kishnani, P.S. Predicting cross-reactive immunological material (CRIM) status in Pompe disease using GAA mutations: Lessons learned from 10 years of clinical laboratory testing experience. *Am. J. Med. Genet. C Semin. Med. Genet.* **2012**, *160*, 40–49. [CrossRef] [PubMed]
10. Chien, Y.H.; Hwu, W.L.; Lee, N.C. Newborn screening: Taiwanese experience. *Ann. Transl. Med.* **2019**, *7*, 281. [CrossRef] [PubMed]
11. Chiang, S.-C.; Chen, P.-W.; Hwu, W.-L.; Lee, A.-J.; Chen, L.-C.; Lee, N.-C.; Chiou, L.-Y.; Chien, Y.-H. Performance of the Four-Plex Tandem Mass Spectrometry Lysosomal Storage Disease Newborn Screening Test: The Necessity of Adding a 2nd Tier Test for Pompe Disease. *Int. J. Neonatal Screen.* **2018**, *4*, 41. [CrossRef]
12. Yang, C.F.; Liu, H.C.; Hsu, T.R.; Tsai, F.C.; Chiang, S.F.; Chiang, C.C.; Ho, H.C.; Lai, C.J.; Yang, T.F.; Chuang, S.Y.; et al. A large-scale nationwide newborn screening program for Pompe disease in Taiwan: Towards effective diagnosis and treatment. *Am. J. Med. Genet. A* **2014**, *164A*, 54–61. [CrossRef] [PubMed]
13. Chien, Y.H.; Lee, N.C.; Chen, P.W.; Yeh, H.Y.; Gelb, M.H.; Chiu, P.C.; Chu, S.Y.; Lee, C.H.; Lee, A.R.; Hwu, W.L. Newborn screening for Morquio disease and other lysosomal storage diseases: Results from the 8-plex assay for 70,000 newborns. *Orphanet J. Rare Dis.* **2020**. [CrossRef] [PubMed]
14. Chien, Y.H.; Chiang, S.C.; Chang, K.L.; Yu, H.H.; Lee, W.I.; Tsai, L.P.; Hsu, L.W.; Hu, M.H.; Hwu, W.L. Incidence of severe combined immunodeficiency through newborn screening in a Chinese population. *J. Formos. Med. Assoc.* **2015**, *114*, 12–16. [CrossRef] [PubMed]
15. Labrousse, P.; Chien, Y.H.; Pomponio, R.J.; Keutzer, J.; Lee, N.C.; Akmaev, V.R.; Scholl, T.; Hwu, W.L. Genetic heterozygosity and pseudodeficiency in the Pompe disease newborn screening pilot program. *Mol. Genet. Metab.* **2010**, *99*, 379–383. [CrossRef] [PubMed]

16. Chien, Y.H.; Hwu, W.L.; Lee, N.C. Pompe disease: Early diagnosis and early treatment make a difference. *Pediatr. Neonatol.* **2013**, *54*, 219–227. [CrossRef] [PubMed]
17. Wang, Z.; Okamoto, P.; Keutzer, J. A new assay for fast, reliable CRIM status determination in infantile-onset Pompe disease. *Mol. Genet. Metab.* **2014**, *111*, 92–100. [CrossRef] [PubMed]

International Journal of
Neonatal Screening

Review

Second Tier Molecular Genetic Testing in Newborn Screening for Pompe Disease: Landscape and Challenges

Laurie D. Smith [1,2], Matthew N. Bainbridge [3,4], Richard B. Parad [5,*] and Arindam Bhattacharjee [2,*]

1   Department of Pediatrics, UNC Hospitals, Chapel Hill, NC 27599, USA; laurie_d_smith@yahoo.com
2   Laboratory Services Division, Baebies, Inc., Durham, NC 27709, USA
3   Codified Genomics, Houston, TX 77004, USA; mbainbridge@rchsd.org
4   Rady Children's Institute for Genomic Medicine, San Diego, CA 92123, USA
5   Department of Pediatric Newborn Medicine, Brigham & Women's Hospital, Harvard Medical School, Boston, MA 02115, USA
*   Correspondence: rparad@bwh.harvard.edu (R.B.P.); abhattac2@gmail.com (A.B.);
     Tel.: +1-617-732-7371 (R.B.P.); +1-978-821-6172 (A.B.)

Received: 17 February 2020; Accepted: 3 April 2020; Published: 5 April 2020

Abstract: Pompe disease (PD) is screened by a two tier newborn screening (NBS) algorithm, the first tier of which is an enzymatic assay performed on newborn dried blood spots (DBS). As first tier enzymatic screening tests have false positive results, an immediate second tier test on the same sample is critical in resolving newborn health status. Two methodologies have been proposed for second tier testing: (a) measurement of enzymatic activities such as of Creatine/Creatinine over alpha-glucosidase ratio, and (b) DNA sequencing (a molecular genetics approach), such as targeted next generation sequencing. (tNGS). In this review, we discuss the tNGS approach, as well as the challenges in providing second tier screening and follow-up care. While tNGS can predict genotype-phenotype effects when known, these advantages may be diminished when the variants are novel, of unknown significance or not discoverable by current test methodologies. Due to the fact that criticisms of screening algorithms that utilize tNGS are based on perceived complexities, including variant detection and interpretation, we clarify the actual limitations and present the rationale that supports optimizing a molecular genetic testing approach with tNGS. Second tier tNGS can benefit clinical decision-making through the use of the initial NBS DBS punch and rapid turn-around time methodology for tNGS, that includes copy number variant analysis, variant effect prediction, and variant 'cut-off' tools for the reduction of false positive results. The availability of DNA sequence data will contribute to the improved understanding of genotype-phenotype associations and application of treatment. The ultimate goal of second tier testing should enable the earliest possible diagnosis for the earliest initiation of the most effective clinical interventions in infants with PD.

Keywords: newborn screening; lysosomal storage diseases; variant cut-off; next generation sequencing; diagnosis; dried blood spots

---

1. Introduction

Newborn screening (NBS) for Pompe disease (PD), utilizes dried blood spots (DBS) to detect deficient alpha-glucosidase (GAA) activity. As part of currently employed NBS algorithms for PD screening in public health laboratories (PHLs), a first tier biochemical test based on MS/MS or digital microfluidics measures the enzyme activity. The disease is suspected if the enzyme activity is below a previously established cut-off value [1]. However, without two tiered or reflex testing, programs using only first tier biochemical assays would yield poor positive predictive values with high false positive rates or risk high false negative rates due to stringent first tier cut-offs [2]. False-positive results on first tier testing of PD create a high referral burden, often related to the prevalence in the population of non-disease causing pseudodeficiency alleles that lead to low enzyme activity without causing PD. Inclusion of a second tier test, on first tier positive NBS samples provides an opportunity to resolve these false-positives prior to reporting, thus avoiding re-contact, needless referrals and creation of parental anxiety. When GAA activity is below established cutoff values, a second tier test can confirm or disprove the diagnosis of PD. Current algorithms applied by NBS programs may perform repeat sampling and redo the GAA enzyme activity test [3]; or use a second tier test (in-house or as a send-out service) such as DNA sequencing that uses traditional DNA Sanger sequencing technique [4], or the newer targeted next generation sequencing (tNGS) method. Recently, a new second tier test marker—the ratio of creatine/creatinine to alpha-glucosidase activity—has also been proposed [5].

From a clinical standpoint, PD has a wide spectrum of phenotypes ranging from early onset with muscle and cardiac involvement (infantile-onset PD; IOPD) to later juvenile or adult onset (later-onset PD; LOPD). LOPD symptomatology may overlap with other neuromuscular disorders, and timely diagnosis is challenging for both forms of the disease. Most early onset cases are symptomatic in some form at birth. Two NBS algorithms for diagnostic confirmation have been developed by a group of international experts on both NBS and PD, the Pompe Disease Newborn Screening Working Group, based on whether DNA sequencing is performed as part of the screening algorithm [6]. Applying the recommendations of either algorithm can lead to a diagnostic characterization as: (a) classic IOPD, (b) "predicted" LOPD, or (c) no disease/not affected/carrier. In both algorithms, a variety of clinical tests are necessary to confirm the diagnosis and generate a treatment plan, including DNA sequencing, since the *GAA* gene variant analysis is essential for confirming the diagnosis and developing a treatment strategy of PD [6]. A challenge to pursuing DNA sequencing as part of NBS algorithms is that many NBS PHLs and clinical referral centers do not perform or have ready access to sequencing resources on premises. When an NBS laboratory does not provide rapid DNA sequencing results through either send-out services or sequencing on premises, the responsibility of obtaining the *GAA* gene sequencing result that is necessary for diagnosis and treatment initiation falls directly on the referral center. Delays in obtaining DNA sequencing results may lead to poorer outcomes or possible loss to follow-up. Loss to follow-up is a concern given the spectrum of phenotypes, many of which present with a delay in onset of symptoms and loss in benefit opportunity of early treatment. If an infant is found to have any form of disease, follow-up with appropriate treatment is necessary and should start early. Recent reports on IOPD continue to shed light on some of the unique challenges care providers face in diagnosing and managing this genetic disease [2,7]. While delays in PD symptom onset and diagnosis are less common in IOPD (under 3–4 months) compared to LOPD, based on the Pompe Registry data [8], the IOPD cases with cross-reactive immunological material (CRIM), negative statuses almost invariably develop high antibody titers to enzyme replacement therapy (ERT). As high antibody titers to ERT reduce the effectiveness of ERT treatment, CRIM-negative patients require immunomodulation therapy prior to initiation of ERT. If CRIM negative IOPDs are not identified in the newborn period, and if immune modulation therapy and ERT treatment are not initiated within days of birth, 100% ventilator free-survival is unlikely, and death may occur within the first two years of life [2,9]. Several Australian IOPD cases that were not identified by NBS, but identified by clinical examination alone have responded poorly to ERT and developed high antibody titres [7]. CRIM-positive IOPDs can also show high antibody titres, which may be predicted based on genotype [10,11]. Rapid recognition of PD at the molecular genetics level, in conjunction with clinical characterization, can guide treatment strategies that may include ERT for IOPD (with or without immune modulation therapy) and LOPD, and may be critical to ensuring the best patient outcomes while preventing irreversible clinical changes [2,12,13]. Given the clinical workup necessary after NBS to confirm a PD diagnosis and direct therapeutic strategy, and the time sensitive nature of the treatment in some PD cases, it may be imperative to also choose a second tier tNGS test that aligns with clinical algorithms and avoids delayed initiation of beneficial PD treatment. Stepwise serial testing between NBS and referral centers, can lead to obtaining DNA results in weeks, delaying initiation of treatment. The typical confirmatory diagnostic DNA sequencing test takes 3–6 weeks and may require reimbursement authorization, and therefore, is inadequate for the prompt genetic testing required for PD. If rapid DNA sequencing testing is included as part of NBS second tier testing, such delays are avoided. Furthermore, such testing may be more equitable for individuals who have no insurance coverage for post-NBS DNA sequencing test. Such decisions, of course, have cost implications for NBS programs.

We therefore also present the rationale for optimizing second tier molecular genetic testing emphasizing (a) fast turn-around time; (b) inclusion of variant effects such as cross-reactive immunological material (CRIM) status, predicted age of symptom onset and heuristics that impact treatment strategies; (c) incorporation of copy number variation and library preparation methods from DBS as part of a tNGS second tier algorithm; and (d) variant prioritization 'cut-off' tools to reduce the number of false positives and to help with PD phenotype prediction. Challenges in addressing health information privacy, policy, regulation, parental consent, and secondary or incidental findings associated with genetic tests are topics beyond the scope of this review.

2. Current Approach to Second Tier and Follow-Up Testing

The Taiwanese PD NBS algorithm measures GAA, neutral α-glucosidase (NAG), and maltase-glucoamylase (MGA) activities in separate fluorometric assays, using commercially available fluorogenic (4-methylumbelliferone) substrates [2] normalized to protein concentrations. These values discriminate between true GAA deficiency-confirmed positives from false-positives. The percentage of acarbose inhibition as a second tier test, for samples with inconclusive results, further improves the performance of testing in this algorithm.

A second tier test marker for NBS of PD recently reported by Tortorelli et al. [5] describes a reduced false-positive rate through the use of a ratio calculated between the creatine/creatinine (Cre/Crn) ratio as the numerator and the activity of acid α-glucosidase (GAA) as the denominator. This ratio is incorporated alongside a post-analytical tool: Collaborative Laboratory Integrated Reports (CLIR; https://clir.mayo.edu) and routinely used in second tier evaluations to address the issue of false positives due to pseudodeficiency.

Because DBS GAA enzyme activity measured in the first tier NBS cannot predict PD phenotype, additional follow-up evaluation using chest X-ray, electrocardiogram (EKG), creatine kinase (CK) levels, pro-B-type natriuretic peptide (pro-BNP) levels, echocardiogram and genotyping or DNA sequencing (to confirm the presence of two pathogenic variants) is required. Western blot analysis of cultured skin fibroblast lysates has been the gold standard for determining cross-reactive immunologic material (CRIM) status [14], although rapid blood-based assays are also available. Other follow-up tests such as urinary Glc4 levels are also recommended [2,6].

3. The Variant Spectrum of Pompe Disease

PD is an autosomal recessive disorder resulting from two pathogenic variants, including multi-exonic deletions, in the GAA gene. The variant spectrum in the *GAA* gene is highly heterogeneous. To date, over 500 pathogenic (P) or likely pathogenic (LP) variants and numerous benign variants and variants of unknown significance (VUS) in the *GAA* gene have been reported (ClinVar, 2017; Erasmus MC University Medical Center, 2017; Aggregation Databases (ExAC, gnomAD), 2017; Leiden Open Variation Database (LOVD), 2018; The Human Genome Mutation Database (HGMD), 2017). A review of *GAA* variants suggests that missense variants are the most frequent molecular genetic cause of PD (~50%), followed by small deletions [8,15]. Variant hotspots are also known, such as the c.-32-13T>G splice site single nucleotide variant and the exon 18 deletion (c.2481+102_2646+31del). These and other ethnic variants are well described [8,15], but require detection by DNA sequencing methods that can identify both single nucleotide variations (SNVs), as well as copy number variations (CNVs, also known as deletion/duplication events), and these are discussed later.

4. Hybrid Capture tNGS as a Second Tier Method

For NBS programs utilizing second tier DNA sequencing, both tNGS on small panels of genes and exome sequencing are considered (see Section 9). Cost and time efficiency needs, along with specimen requirements and future expansion needs, make DNA hybrid-capture based sequencing protocols for tNGS ideal for PD screening/diagnostic algorithms (see Section 5 below for additional considerations). A reduced time through the pipeline due to a more efficient process means shorter time to results and needed treatment for affected infants. A hybrid-capture based tNGS workflow [16], as outlined in Figure 1, may be performed on 1–2 newborn DBS punches, with end-to-end testing time from sample receipt to initiating reports of ~35 h (Figure 1a). In our experience, this approach used for the NBS second tier testing of PD and other lysosomal storage disorders leads to the return of results in less than a week. Reagents used are typically 'kit-based' and the processing protocol may be optimized specifically as needed. Most send-out DNA sequencing services typically have a turn-around time of several weeks, as well as referral and sample redraw, making this process challenging for the rapidity needed in NBS protocols. In our experience of processing second tier tNGS for several US PHLs, we have been able to report variant information in seven days, and reduce the referral burden in some cases by 70% (reports that did not have any P/LP variant or had exclusively pseudodeficiency variants). These reports, therefore, were capable of providing carrier status, CRIM prediction, and severity of the known reportable variants, which would not otherwise be possible.

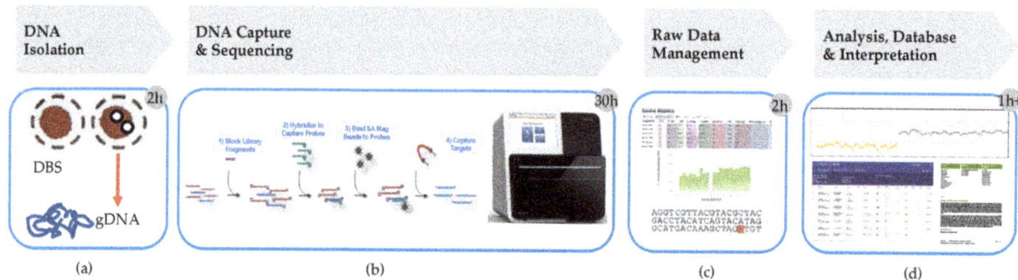

Figure 1. Hybrid Capture-Sequencing tNGS Workflow. Many components of the workflow may be automated: (**a**) DNA Isolation—This step in the workflow may be performed with two 3.2 mm DBS punches, takes 2 h (h) to perform and is an essential requirement of second tier testing as most tNGS protocols call for 2–10 mL of blood; (**b**) DNA hybrid Capture and Sequencing—DNA capture requires library preparation and bait based DNA hybridization-based capture of target regions, which is followed by DNA sequencing (Illumina MiniSeq). (**c**) Raw data management—Involves moving data across secure compliant environments for storage, processing and linking into databases; (**d**) Analysis, database and interpretation—Involves analysis for single nucleotide variants and copy number variant callers, comparative analysis with other database annotations, interpretation and reporting. This overall process takes ~35 h. Adapted from Bhattacharjee et al. [16].

DNA library preparation involves sonication or a kit-based enzymatic DNA fragmentation step (amenable to automation and high throughput) followed by subsequent adaptor ligation/indexing. Indexed DNA fragments are then hybridized to custom DNA enrichment probes in order to target the specific genes on the panel (Figure 1b). Such DNA probes are available from various commercial sources and may be custom designed to target 100–200 genes. Sequencing may be performed on several sequencing platforms. The Illumina MiniSeq platform shown in the example (Figure 1b), takes around 14h to complete a run and can parallel process 20 samples per run. The bioinformatics workflow (raw data management) is processed mostly in parallel and can take an additional 2 h. Custom QC metrics provide insight into sequence quality for each sample as well as coverage gap analysis (Figure 1c). A custom copy number variation (CNV) caller can be used on the same sequence dataset to identify CNVs (Figure 1d, Figure 2a). Strict quality metrics prevent CNV calling on samples with low coverage. An in-house variant database with a well-established analysis platform, such as from Opal™ (Fabric Genomics), can appropriately annotate the data set in real-time and help generate sequencing reports more efficiently. Automated variant scoring allows a reduction in time for data interpretation, and thus drives higher throughput.

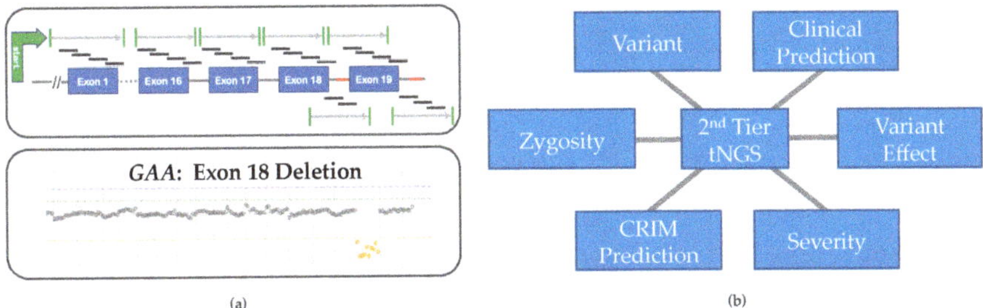

Figure 2. Second tier Testing features: (**a**) Top panel shows how *GAA* gene structure is targeted for sequencing with overlapping single stranded DNA or RNA probes of 60–200 nucleotides in target exonic regions (blue rectangles), adjoining splice junctions (marked in green), and known hotspots (red). The bottom panel shows the results of a in-house algorithm that integrates detection of copy number variation (CNV) or del/dup events in Pompe disease, such as the *GAA* exon 18 deletion; (**b**) 2nd tier tNGS provides variant (genotype) information for better clinical predictions and management of care. Pseudodeficient alleles, variant effect, and cross-reactive immunologic material (CRIM) prediction can be elucidated from previously observed data [8,14].

5. Ideal PD Second Tier tNGS Testing Features

5.1. Timeliness

The US Federal and State Public Health programs emphasize the importance of timeliness of NBS, setting benchmarks for the efficient collection, transportation, testing, and reporting of results. First and second tier tests are combined to optimize detection of cases at a false positive rate as low as possible and keep overall costs of the combined tests low. Upon referral, necessary diagnostic tests are done. For many diseases that progress slowly, the split in the burden between screening and diagnostic phases may be sufficient. However, this tiered/serial testing and patient recall can cause delays in time-sensitive PD. For most of the LOPD cases that progress slowly, this approach would be sufficient without causing harm. However, for suspected IOPD cases, time to treatment initiation within two weeks is critical. Therefore, timely second tier tNGS reporting in IOPD may be the only option for avoiding diagnostic and treatment delays and avoiding a negative impact on quality of life and potentially on lifespan [7,13]. The treating clinician, in the case of classic IOPD, would need DNA sequencing data in the diagnostic-phase to determine treatment-associated predictive sequence information (CRIM status and phenotype onset predictions) prior to initiating enzyme replacement therapy [2,6]. Although controversial, suspected LOPD patients identified by NBS second tier tNGS may also benefit from follow-up and earlier intervention [2,13,17]. Rapid return of results is a frequent claim by providers of DNA sequencing for diagnostic testing. However, the DNA sequencing turnaround time is generally inadequate for early onset PD (minimum industry standard is three weeks' turnaround). Such inefficiencies make utilization of most commercial resources untenable for the required timeframe of treatment following the diagnostic phase. Second tier tNGS workflow services, if performed end-to-end, may be completed in ~35 h, and reports can be finalized consistently within 5–7 days of receiving the sample. If NBS programs routinely perform tNGS on premises, or use DNA sequencing services that return results within 7 days, then overall reporting and treatment initiation may occur within 10 days, benefiting classic IOPD or CRIM-negative IOPD cases. Starting ERT within this timeframe for CRIM-negative IOPD, and before the immune system has matured, might also enable natural immune tolerance [7]. Omitting the tNGS assay as the second tier PD NBS test and putting the burden on the follow-up referral centers to obtain a molecular diagnostic result, while providing a faster NBS turn-around time may delay treatment initiation. Due to the fact that the optimal effectiveness of ERT in IOPD hinges on the prompt identification of two *GAA* pathogenic variants in-trans [6], a second tier tNGS assay may actually be beneficial by shortening the time to treatment initiation. Identification of one or two pathogenic variants (without *cis* or *trans* determination), combined with a low GAA activity is sufficient to consider need for further confirmatory testing.

5.2. Prediction of Genotype-Phenotype Effects

These topics have been well covered by other reviews [8,15], and therefore only discussed in the context of second tier testing algorithm for PD NBS (Figure 2). In PD, the genotype is generally fully penetrant; IOPD phenotypes especially demonstrate limited heterogeneity. A pseudodeficiency allele is always a pseudodeficiency allele as it causes low measured GAA activity that does not change from individual to individual. However, predicted CRIM status may not exactly correlate as observed in some patients with Western blot analysis of cultured skin fibroblast lysates or for splice variants and may require additional studies [14]. Identification of variants in the *GAA* gene provides valuable information for determining variant-phenotypes, establishing genotype–phenotype correlations, confirming diagnosis or carrier status and counseling about the significance of these findings [6]. Measuring first tier GAA activity, genotyping, and determination of CRIM status from genotype (important for phenotype onset prediction) is necessary to start therapy [14,18].

5.3. Prediction of CRIM Status

In general, for babies with IOPD who have cardiac involvement, treatment with recombinant human GAA (rhGAA) is initiated immediately after confirmation of GAA deficiency and positive CRIM status, either by genotyping prediction or by western blot [2,6]. ERT is started after the cardiac involvement is confirmed, unless immunomodulation therapy to prevent anti-GAA antibodies production is planned. For infants without cardiac involvement, close follow up is needed and ERT treatment is delayed until symptoms appear. As confirmatory biochemical tests and assays are not readily available for PD, and additional testing may delay initiation of therapy, sequence variant identification may be useful to predict CRIM status, thereby directing treatment decisions [14,18]. The presence of two nonsense or frameshift-termination variants is a good predictor of CRIM status, typically resulting in CRIM-negative status, unless the premature stop codon is in the last exon of the gene [14].

5.4. Copy Number Variation (CNV) and False Negative Risk

Due to the fact that CNV events such as those reported in *GAA* exon 18 [8,15] and other heterogeneous CNVs comprise a significant fraction of pathogenic PD variants, a PD NBS second tier tNGS assay must integrate detection of CNV events. While data from targeted small gene panels are scarce, recent reports of PD whole exome sequences provide insight into CNVs. Mori et al. [19] evaluated whole exome sequencing to identify infantile- and late-onset PD ($n = 93$), concluding that some pathogenic variants may have been missed, since the variant pipeline used did not identify variant hotspots such as the exon 18 deletion (Figure 2a), and c.-32–13 single nucleotide variant. A sensitivity of near 100% can be achieved if issues with custom design, gap filling, improving >20X sequence coverage and deletion duplication coverage are addressed.

5.5. False Positives Resolved by Second Tier tNGS Testing

The variant burden and its effect on an individual is unknown at birth, and phenotype characterized by biochemical activity alone may be imprecise. Low GAA enzyme activity detected by first tier assays may identify pseudodeficiency cases and carriers of a single pathogenic *GAA* variant as a false positive, thus enzyme activity results alone are not definitive and cannot differentiate between true PD and newborns who carry pseudodeficiency alleles or sometimes PD carriers (P/LP variants). Such cases can be distinguished by genetic testing.

5.6. Other Considerations

Occasionally it may be difficult to resolve a false positive with second tier tNGS. Hypothetically, if a sample has a low enzyme activity value on first tier testing and there are no findings on second tier tNGS, the etiology of the low enzyme activity has not been identified and the result can be called negative in screening, or an additional biochemical test can be done. The etiology of discordant results may involve sample mix up or other technical issues. A sample mix up can sometimes be addressed by comparing imputed sex from tNGS data to that reported on accession data. Monitoring of additional DBS enzyme activity values may also indicate false positive risk. For example, if some additional enzyme activity values are all low, it may suggest issues with stability, storage or transportation. As mentioned, new post-analytical tools, such as Collaborative Laboratory Integrated Reports (CLIR; https://clir.mayo.edu) and second tier biochemical testing are becoming available to address these issues [5,17]. A tNGS false negative result when the first tier has low enzyme value is very unlikely given the recessive nature of PD, where the presence of at least two causal variants in trans is expected. We discuss below the post-analysis of tNGS data, including statistical tools to predict and reduce the number of false positives and understand PD phenotypes.

6. Genome Scale Data and Its Impact on PD Screening

We evaluated how information in public genomic databases would inform and impact PD NBS that utilized second tier tNGS. In order to identify potential pathogenic variants for the detection of PD through NBS, we reviewed *GAA* allele frequencies in the gnomAD database. Based on gnomAD and ClinVar, pathogenic and likely pathogenic (P/LP) variants have a total allele frequency of 0.635%. Assuming these alleles segregate independently and are fully penetrant, the probability of inheriting two such alleles is the square of 0.635 or 0.004%. This roughly equates to an expected incidence of 1/24,780 births, which is in the expected range for PD (1/14,000 to 1/100,000). This means that a significant proportion of the variant alleles that cause PD are known and can be automatically identified in a high-throughput fashion.

For the sake of timeliness, it is critical to have a tNGS assay that can identify all genetic variations in a single test. To assemble relevant variants for the diagnosis of PD through NBS, we reviewed alleles in the gnomAD database for potential CRIM status. GnomAD's *GAA* sequences from 141,456 individuals revealed 213 (0.15%) predicted LOF variants (pLOF) at 82 unique sites, of which approximately 50% were frameshift termination and 25% each were splice-site and stop-gain variants. None of these mutations were present in a homozygous state and although an additional mutation may be present in trans; it is more likely that these individuals are presumed carriers. Based on gnomAD data, the carrier frequency of unique pLOF variants is 0.15% (213/141,456) in the population, which is 4-fold less than P/LP variants in gnomAD. Estimates of carriers in general uncovered by NBS is a fraction (0.007–0.013%) of that observed frequency in gnomAD [17]. Although some may be, most PD carriers will not be detected by NBS program based on pLoF or P/LP variant frequencies in gnomAD. While reporting on carrier status is not the aim of most NBS programs (most try to avoid detecting carriers), parents of affected PD patients are obligate carriers. The risk for two carrier parents to have an affected child is 25% with each pregnancy. Thus, the mode and impact of reporting carrier status to parents after detection through a PD tNGS second tier NBS algorithm require careful consideration. Other NBS disorders deal with this issue routinely (e.g., cystic fibrosis carriers via a DNA based second tier, or sickle cell anemia via electrophoresis).

A family history of negative PD screens without carrier information gives families a false sense of security, as reported in the Australian PD cases [7]. A significant fraction of FPs will be due to recurrent carriers or pseudodeficiencies with low enzyme values, and a likely source of anxiety or rightful concern in future newborn births in those families. Thus, a delay in relaying carrier or pseudodeficiency information may have secondary consequences that may need to be considered. Recently, the BabySeq Project (a part of NSIGHT), evaluated newborn diseases using genome scale sequencing in a randomized clinical trial pilot. The cohort studied included both well newborns and those admitted to a neonatal intensive care unit [20,21]. In a cohort of 159 newborns that underwent genome scale NGS based whole exome sequencing (WES), one carrier of a pathogenic *GAA* variant was identified. As part of that study protocol, the in-person genetic counseling was provided to the family.

We also queried 14,821 WES individuals referred for clinical genetic testing to determine a 'hit rate' for *GAA*, generating a list of reportable variants by curating pathogenic and likely pathogenic variants from reputable sources in ClinVar and from manual literature curation. These estimates are useful in understanding PD causal variant frequencies and database quality. This curation resulted in over 100 *GAA* reportable variants. For each case, the exome variant file (variant call format (VCF) file) was scanned through our database and scored as to whether that sample was identified ("hit"). Only one homozygous hit (allele fraction > 90%) was observed for *GAA*. Due to the fact that testing was performed on single samples, it was not possible to know if multiple variants were in cis or trans. Thus, we defined a compound heterozygous hit as having two or more variants in a single gene, in a single sample, from our reportable list of variants. Only two cases had reportable variants in *GAA* and thus the potential "hits" for the referred population was 2/14,821 (0.014%), a value that was consistent with the incidence of PD. These types of analyses have several limitations: (a) no ethnicity control was available, as ethnicity of patients was not known, and therefore impact on actual screening could be variable; (b) consanguinity of patients was not known; (c) patients in the dataset were likely depleted of traditional PD; (d) it was not possible to distinguish the cis and trans configuration of multiple hits in a single individual; (e) this analysis did not take into account novel variants, nor did it take into account copy number variants; (f), variant calls were affected by coverage of the gene/base and particularities of the caller. Thus, while it was not possible to perfectly match the false positive /false negative rate of the caller, this effect was likely small.

7. Variant 'Cut off' for PD

To reduce the burden of interpreting variants of uncertain significance (VUS), we considered a 'cut off' model to filter variants that are too common in population to cause disease. We estimated the maximum credible allele frequency (MCAF), using a statistical framework that was previously reported for a recessive inheritance model by Wiffin et al. [22] Variants that occur more frequently than the MCAF cutoff should not be considered as a causative variant for the disease.

The equation to calculate MCAF is as follows (1):

$$\text{MCAF} = \text{sqrt}(v) \times c \times \text{sqrt}(g) \times 1/\text{sqrt}(p) \tag{1}$$

where c is the maximum allelic contribution, which represents the proportion of cases that are attributable to the gene that is attributable to an individual variant, g is the maximum genetic contribution representing the proportion of all cases that are due to the gene, v is the prevalence and p is the penetrance. This value can be further refined by assuming that the most common causative alleles are known.

For the *GAA* gene, we identified common alleles that were both expertly curated and present in ClinVar. Furthermore, we established a set of trusted ClinVar submissions (multiple sites with no conflicts (GeneDx and Invitae)) and identified all P/LP variants. From this we identified the most common pathogenic variants that contribute to the disease (ClinVar and trusted submitters). We then made the assumption that once the most common variants are accounted for, the maximum allelic contribution of any additional variant is 5%. We use 0.000025 (1/40,000) as the incidence of high-risk PD, 1 as the penetrance and assume all causes of PD are attributable to variants in *GAA*. The calculation for MCAF is then: sqrt(0.000025) × 0.05 × sqrt(1) × 1/sqrt(1) or 0.00025. We then compare this to the maximum MAF in manually curated or trusted submitter (Table 1). This value can then be used to remove likely spurious entries in ClinVar from further consideration. Furthermore, novel variants (VUSes) with MAFs above this cut-off can likely be discarded as likely benign.

Table 1. Maximum Credible Allele Frequency (MCAF): Allelic Contribution, Penetrance, MAF cutoff, pathogenic variants in ClinVar from Trusted submitters and final MAF cutoff.

Prevalence	Max. Allelic Contribution	Penetrance	MCAF (Whiffin et al. [22])	Max. MAF in ClinVar	Final Cut Off
0.000025	0.05	1	0.00025	0.00358	0.00025

8. Follow-Up Infrastructure for Families, Screen Positive Infants and Carrier Status

Some of the most difficult issues generated by NBS for PD will be the follow-up care, for not only the infants who screen positive, but also for the parents and families. As of 2017, the Pompe Disease Newborn Screening Working Group has proposed options for including DNA sequencing as part of the screening algorithm or follow-up [6]. The Pompe Registry, a long-term, multinational observational program (NCT002314000), designed to improve understanding of the natural history and outcomes of patients with PD, started in 2004, and is sponsored and administered by Sanofi Genzyme (Cambridge, MA), the pharmaceutical company that markets PD ERT. The Pompe Registry has *GAA* variants and phenotypes for 1079 patients [8]. Despite the existence of this network, challenges surround the referral of those patients to appropriate care centers, to ensure that up to date care guidelines that reflect the current standards of the PD NBS community are adhered to. Given that current disease modifying treatments for PD may depend on knowing the specific genetic defect, it is critical and ethically required, that all screen positive infants have timely access to genetic testing, as either part of NBS or as referral within the timeframe. Testing should define the specific variants and access to appropriate follow up clinical expertise, so that the appropriate treatment is provided. Future therapies, once approved, such as chaperones, c.-32-splice switching antisense oligonucleotides and AAV-directed gene therapies will need variant information.

Testing parents to determine which parent carries which PD variant is necessary to provide appropriate counseling regarding future reproductive risk for family members. Knowing familial variants for PD can be useful for reproductive planning and the avoidance of recurrence, but is not the primary goal of NBS. Most guidelines recommend against carrier testing of minors and minor siblings, as there is no medical benefit until reaching reproductive age and age of consent [23–25].

Some screen positive infants will have an early-onset form of the disease while others will develop LOPD. Knowing the precise genetic defect in a screen positive IOPD case is critical for the prediction of the long-term clinical outcome of ERT and the development of antibodies against the infused enzyme. An analysis of CRIM status for every IOPD patient within the first two weeks of life is essential. For this reason, sequence analysis of the GAA gene may be justified as a tiered testing step for PD NBS programs or a rapid turnaround confirmatory test for the referral center. Some first tier screen positive newborns may have resolution on a tNGS second tier test, as they may carry pseudodeficiency alleles or a variant burden, that are otherwise low risk. For parents of screen positive infants with identified variant(s), familial testing is relatively simple and may be considered in reproductive planning. Normal GAA enzyme activity, or an above threshold value on a PD NBS, does not confirm a non-carrier status. It is possible that a pathogenic variant may have been missed or that a phenotype may manifest later. The potential therefore exists for misconstruing a negative PD NBS. Further studies must evaluate whether all infants identified with disease or as carriers have received adequate follow-up, including diagnostic confirmation after a positive screen, referral to appropriate clinical care centers and delivery of best practice treatment and management across the lifespan. State NBS programs must have a short-term follow-up plan in place to ensure tracking of screen positive newborns for the receipt of proper referral to appropriate care centers for diagnosis and treatment. Ideally, this short-term follow-up program should have access to resources that include: (a) *GAA* sequencing availability or referral for rapid turnaround DNA sequencing for screen positive infants; (b) the ability to provide cascade genetic testing of both family and extended family members; and (c) the ability to carry out all necessary confirmatory testing. Furthermore, referral sites should be able to appropriately triage infants found to have incidental findings like other overlapping neuromuscular diseases, such as Limb Girdle muscular dystrophy. Since it is recommended that CK levels be determined in all infants in whom 1st and 2nd tier testing is suspicious of PD, there will likely be utility in determining baseline CK levels from DBS at the time of initial enzyme testing. This would be helpful not only for confirming the diagnosis and establishing a baseline level prior to starting treatment, but also allowing for early identification of other neuromuscular disorders if there is persistence of hyperCKemia, despite negative testing for PD. As with all genetic testing, identifying and implementing appropriate educational programs is key. Improving health care provider and parental genetic literacy about genetic disorders is required.

9. Current and Future Utilization of tNGS Testing

PD was added to the U.S. Recommended Universal Screening Panel (RUSP) in 2015. Targeted NGS could reduce the treatment delay for those identified by PD screening by allowing analysis, interpretation, and appropriate reporting of healthcare related information in a timely manner. Already several PHLs in the USA have adopted second tier tNGS in-house or as a service [26,27]. New York and California PHLs have strong DNA sequencing second tier programs including tNGS for PD and other NBS disorders. Several other NBS PHLs use 'send-out second tier tNGS services' for PD. Given developments in genome scale technologies, such as WES or whole genome sequencing (WGS), the utilization of tNGS may be co-opted for needs in second tier testing of PD. Recently, the Utah PHL has started implementing a second tier exome sequencing protocol for genes associated with newborn screening abnormalities [28]. Furthermore, tNGS based testing can avoid constraints or complicating factors associated with biochemical testing, such as the infant's gestational age at birth, transfusion status, age at sample collection, need for repeat sampling (rescreens or redraws), and metabolic and feeding states. The NIH funded NSIGHT consortium (Newborn Sequencing in Genomic Medicine and Public Health) evaluated newborn diseases using genome scale sequencing in randomized clinical trials [21,22,29]. This is an exciting time to be at the forefront of applying genomic information to rapidly identifying and treating genetic disorders such as PD. We envision that population level screening as first tier WES or WGS, or even first tier small gene panel based tNGS testing, is unlikely to be the avenue by which PD may be identified in the near term, due to considerations such as cost and complexity. However, with improvements in technology and a decrease in cost, this may likely be the NBS of the future.

10. Conclusions

In summary, molecular based NGS testing approaches are suitable for NBS of PD, and may be used in algorithms that include either a tandem second tier test or in a contingent fashion. The primary goal of NBS is identifying patients who can be treated to establish significant health gain. Secondary goals, such as shortening the diagnostic odyssey, identifying carriers and providing information for reproductive options are of lesser concern. However, as PD is a spectrum disease simply providing biochemical screen positive information alone delays treatment initiation in those phenotypes with rapid disease progression, which impacts the ultimate outcome. PD has both pseudodeficiencies as well as early-onset and late-onset clinical phenotypes, presenting a special challenge to families and medical follow-up centers. Early genotype information in PD management allows for prompt treatment initiation and the potential for better clinical outcomes, especially in infantile onset disease [2,3,9,12,13]. Considering the recommended diagnostic algorithm [2,6], of which DNA sequencing is a part, the use of second tier tNGS sequencing is only logical, irrespective of whether the PHL yet has sequencing capabilities, since NBS may be the only way to provide molecular genetic information for early determination and treatment of PD in an equitable manner. Several PHLs including New York and California have already introduced DNA sequencing in their PD NBS algorithms. While second tier biochemial tests are capable of PD identification, second tier tNGS test for PD can provide rapid and precise information on highly penetrant recurrent pathogenic variants, distinguish pseudodeficiency alleles with lower biochemical values that are false-positives, provide clues for CRIM status, and identify variants associated with IOPD or LOPD phenotypes. A combination of early detection, close monitoring, and early ERT is likely to be beneficial to LOPD patients, but additional data are needed. Consistent with the original intent of NBS, PD must be considered a time critical condition, for which all PD NBS results including second tier should be provided as early as possible. Given the ever-increasing population-based variation outcome data for PD [30], implementation of tNGS second tier testing of PD from a DBS [16], is both feasible and sufficient in the required time frame for NBS and follow up.

Funding: A.B. was partly supported by National Institutes of Health (NICHD) Grant (R43HD094543).

Acknowledgments: We thank Rebecca Kitchener and Viren Amin for the figures, technical details and review of the manuscript.

Conflicts of Interest: R.B.P. and M.N.B. declare no conflict of interest. L.D.S. is an independent consultant for Baebies, Inc., Durham, NC. A.B. is an employee of Baebies, Inc., Durham, NC.

References

1. Gelb, M.; Lukacs, Z.; Ranieri, E.; Schielen, P. Newborn Screening for Lysosomal Storage Disorders: Methodologies for Measurement of Enzymatic Activities in Dried Blood Spots. *Int. J. Neonatal Screen.* **2018**, *5*, 1. [CrossRef] [PubMed]

2. Chien, Y.-H.; Hwu, W.-L.; Lee, N.-C. Newborn Screening: Taiwanese Experience. *Ann. Transl. Med.* **2019**, *7*, 281. [CrossRef] [PubMed]

3. Yang, C.-F.; Liu, H.-C.; Hsu, T.-R.; Tsai, F.-C.; Chiang, S.-F.; Chiang, C.-C.; Ho, H.-C.; Lai, C.-J.; Yang, T.-F.; Chuang, S.-Y.; et al. A Large-Scale Nationwide Newborn Screening Program for Pompe Disease in Taiwan: Towards Effective Diagnosis and Treatment. *Am. J. Med Genet. Part A* **2013**, *164*, 54–61. [CrossRef]

4. Wasserstein, M.P.; Caggana, M.; Bailey, S.M.; Desnick, R.J.; Edelmann, L.; Estrella, L.; Holzman, I.; Kelly, N.R.; Kornreich, R.; Kupchik, S.G.; et al. The New York Pilot Newborn Screening Program for Lysosomal Storage Diseases: Report of the First 65,000 Infants. *Genet. Med.* **2018**, *21*, 631–640. [CrossRef] [PubMed]

5. Tortorelli, S.; Eckerman, J.S.; Orsini, J.J.; Stevens, C.; Hart, J.; Hall, P.L.; Alexander, J.J.; Gavrilov, D.; Oglesbee, D.; Raymond, K.; et al. Moonlighting Newborn Screening Markers: The Incidental Discovery of a Second-Tier Test for Pompe Disease. *Genet. Med.* **2017**, *20*, 840–846. [CrossRef] [PubMed]

6. Burton, B.K.; Kronn, D.F.; Hwu, W.-L.; Kishnani, P.S. The Initial Evaluation of Patients after Positive Newborn Screening: Recommended Algorithms Leading to a Confirmed Diagnosis of Pompe Disease. *Pediatrics* **2017**, *140*, S14–S23. [CrossRef] [PubMed]

7. Saich, R.; Brown, R.; Collicoat, M.; Jenner, C.; Primmer, J.; Clancy, B.; Holland, T.; Krinks, S. Is Newborn Screening the Ultimate Strategy to Reduce Diagnostic Delays in Pompe Disease? The Parent and Patient Perspective. *Int. J. Neonatal Screen.* **2020**, *6*, 1. [CrossRef]

8. Reuser, A.J.J.; Ploeg, A.T.; Chien, Y.H.; Llerena, J.; Abbott, M.A.; Clemens, P.R.; Kimonis, V.E.; Leslie, N.; Maruti, S.S.; Sanson, B.J.; et al. On Behalf of the Pompe Registry Sit. GAA Variants and Phenotypes among 1079 Patients with Pompe Disease: Data from the Pompe Registry. *Hum. Mutat.* **2019**, *40*, 2146–2164. [CrossRef]

9. Owens, P.; Wong, M.; Bhattacharya, K.; Ellaway, C. Infantile-Onset Pompe Disease: A Case Series Highlighting Early Clinical Features, Spectrum of Disease Severity and Treatment Response. *J. Paediatr. Child Health* **2018**, *54*, 1255–1261. [CrossRef]

10. Desai, A.K.; Li, C.; Rosenberg, A.S.; Kishnani, P.S. Immunological Challenges and Approaches to Immunomodulation in Pompe Disease: A Literature Review. *Ann. Transl. Med.* **2019**, *7*, 285. [CrossRef]

11. Groot, A.D.; Kazi, Z.; Martin, R.; Terry, F.; Desai, A.; Martin, W.; Kishnani, P. HLA- and Genotype-Based Risk Assessment Model to Identify Infantile Onset Pompe Disease Patients at High-Risk of Developing Significant Anti-Drug Antibodies (ADA). *Clin. Immunol.* **2019**, *200*, 66–70. [CrossRef] [PubMed]

12. Yang, C.-F.; Chu, T.-H.; Huang, L.-Y.; Liao, H.-C.; Soong, W.-J.; Niu, D.-M. AB028. Very Early Treatment for Infantile-Onset Pompe Disease Contributes to Better Outcomes: 10-Year Experience in One Institute. *Ann. Transl. Med.* **2017**, *169*, 174–180. [CrossRef]

13. Chien, Y.-H.; Lee, N.-C.; Chen, C.-A.; Tsai, F.-J.; Tsai, W.-H.; Shieh, J.-Y.; Huang, H.-J.; Hsu, W.-C.; Tsai, T.-H.; Hwu, W.-L. Long-Term Prognosis of Patients with Infantile-Onset Pompe Disease Diagnosed by Newborn Screening and Treated since Birth. *J. Pediatr.* **2015**, *166*, 985–991. [CrossRef] [PubMed]

14. Bali, D.; Goldstein, J.; Banugaria, S.; Dai, J.; Mackey, J.; Rehder, C.; Kishnani, P. Predicting Cross Reactive Immunological Material (CRIM) Status in Pompe Disease Using GAA Mutations: Lessons Learned from 10 Years of Clinical Laboratory Testing Experience. In *American Journal of Medical Genetics Part C: Seminars in Medical Genetics*; John Wiley & Sons: Hoboken, NJ, USA, 2012.

15. Peruzzo, P.; Pavan, E.; Dardis, A. Molecular Genetics of Pompe Disease: A Comprehensive Overview. *Ann. Transl. Med.* **2019**, *7*, 278. [CrossRef]

16. Bhattacharjee, A.; Sokolsky, T.; Wyman, S.K.; Reese, M.G.; Puffenberger, E.; Strauss, K.; Morton, H.; Parad, R.B.; Naylor, E.W. Development of DNA Confirmatory and High-Risk Diagnostic Testing for Newborns Using Targeted Next-Generation DNA Sequencing. *Genet. Med.* **2014**, *17*, 337–347. [CrossRef]

17. Millington, D.; Bali, D. Current State of the Art of Newborn Screening for Lysosomal Storage Disorders. *Int. J. Neonatal Screen.* **2018**, *4*, 24. [CrossRef]

18. Kishnani, P.S.; Goldenberg, P.C.; Dearmey, S.L.; Heller, J.; Benjamin, D.; Young, S.; Bali, D.; Smith, S.A.; Li, J.S.; Mandel, H.; et al. Cross-Reactive Immunologic Material Status Affects Treatment Outcomes in Pompe Disease Infants. *Mol. Genet. Metab.* **2010**, *99*, 26–33. [CrossRef]

19. Mori, M.; Haskell, G.; Kazi, Z.; Zhu, X.; Dearmey, S.M.; Goldstein, J.L.; Bali, D.; Rehder, C.; Cirulli, E.T.; Kishnani, P.S. Sensitivity of Whole Exome Sequencing in Detecting Infantile- and Late-Onset Pompe Disease. *Mol. Genet. Metab.* **2017**, *122*, 189–197. [CrossRef]

20. Holm, I.A.; Agrawal, P.B.; Ceyhan-Birsoy, O.; Christensen, K.D.; Fayer, S.; Frankel, L.A.; Genetti, C.A.; Krier, J.B.; Lamay, R.C.; Levy, H.L.; et al. The BabySeq Project: Implementing Genomic Sequencing in Newborns. *BMC Pediatr.* **2018**, *18*. [CrossRef]

21. Ceyhan-Birsoy, O.; Machini, K.; Lebo, M.S.; Yu, T.W.; Agrawal, P.B.; Parad, R.B.; Holm, I.A.; Mcguire, A.; Green, R.C.; Beggs, A.H.; et al. A Curated Gene List for Reporting Results of Newborn Genomic Sequencing. *Genet. Med.* **2017**, *19*, 809–818. [CrossRef]

22. Whiffin, N.; Minikel, E.; Walsh, R.; O'Donnell-Luria, A.H.; Karczewski, K.; Ing, A.Y.; Barton, P.J.R.; Funke, B.; Cook, S.A.; Macarthur, D.; et al. Using High-Resolution Variant Frequencies to Empower Clinical Genome Interpretation. *Genet. Med.* **2017**, *19*, 1151–1158. [CrossRef] [PubMed]

23. Botkin, J.R.; Belmont, J.W.; Berg, J.S.; Berkman, B.E.; Bombard, Y.; Holm, I.A.; Levy, H.P.; Ormond, K.E.; Saal, H.M.; Spinner, N.B.; et al. Points to Consider: Ethical, Legal, and Psychosocial Implications of Genetic Testing in Children and Adolescents. *Am. J. Hum. Genet.* **2015**, *97*, 501. [CrossRef]

24. *Report on the Genetic Testing of Children 2010*; British Society for Human Genetics: Birmingham, UK, 2010.

25. Ethical and Policy Issues in Genetic Testing and Screening of Children. *Pediatrics* **2013**, *131*, 620–622. [CrossRef] [PubMed]

26. Tang, H.; Feuchtbaum, L.; Sciortino, S.; Matteson, J.; Mathur, D.; Bishop, T.; Olney, R.S. The First Year Experience of Newborn Screening for Pompe Disease in California. *Int. J. Neonatal Screen.* **2020**, *6*, 9. [CrossRef]

27. Burton, B.K.; Charrow, J.; Hoganson, G.E.; Fleischer, J.; Grange, D.K.; Braddock, S.R.; Hitchins, L.; Hickey, R.; Christensen, K.M.; Groeppner, D.; et al. Newborn Screening for Pompe Disease in Illinois: Experience with 684,290 Infants. *Int. J. Neonatal Screen.* **2020**, *6*, 4. [CrossRef]

28. Ruiz-Schultz, N.; Oakson, K.; Jones, D.; Rindler, M.; Hart, K.; Rohrwasser, A. Targeted Second-Tier Confirmatory Next Generation Sequencing Newborn Screening Pipeline. In *Poster Abstracts 2019*; Newborn Screening & Genetic Testing Symposium: Chicago, IL, USA, 7–10 April 2019.

29. Milko, L.V.; Odaniel, J.M.; Decristo, D.M.; Crowley, S.B.; Foreman, A.K.M.; Wallace, K.E.; Mollison, L.F.; Strande, N.T.; Girnary, Z.S.; Boshe, L.J.; et al. An Age-Based Framework for Evaluating Genome-Scale Sequencing Results in Newborn Screening. *J. Pediatr.* **2019**, *209*, 68–76. [CrossRef]

30. Bergsma, A.J.; In't Groen, S.L.; van den Dorpel, J.J.; van den Hout, H.J.; van der Beek, N.A.; Schoser, B.; Toscano, A.; Musumeci, O.; Bembi, B.; Dardis, A.; et al. A Genetic Modifier of Symptom Onset in Pompe Disease. *EBioMedicine* **2019**, *43*, 553–561. [CrossRef]

International Journal of
Neonatal Screening

Article

Lessons Learned from Pompe Disease Newborn Screening and Follow-up

Tracy L. Klug [1,*], Lori B. Swartz [1], Jon Washburn [2], Candice Brannen [2] and Jami L. Kiesling [1]

[1] Missouri Department of Health and Senior Services, P.O. Box 570, Jefferson City, MO 65102-0570, USA
[2] Baebies, Inc., P.O. Box 14403, Durham, NC 27709, USA
* Correspondence: tracy.klug@health.mo.gov

Received: 21 January 2020; Accepted: 11 February 2020; Published: 14 February 2020

Abstract: In 2015, Pompe disease became the first lysosomal storage disorder to be recommended for universal newborn screening by the Secretary of the U.S. Department of Health and Human Services. Newborn screening for Pompe has been implemented in 20 states and several countries across the world. The rates of later-onset disease phenotypes for Pompe and pseudodeficiency alleles are higher than initially anticipated, and these factors must be considered during Pompe disease newborn screening. This report presents an overview of six years of data from the Missouri State Public Health Laboratory for Pompe disease newborn screening and follow-up.

Keywords: Pompe disease; newborn screening; follow-up; pseudodeficiency

---

1. Introduction

Pompe disease, also called glycogen storage disease type II, is a rare genetic disorder in which variants in the glucosidase alpha acid gene (gene abbreviation: *GAA*) result in low levels of acid alpha-glucosidase enzyme (enzyme abbreviation: GAA) and consequent accumulation of glycogen in various tissues of the body [1,2]. The build-up of glycogen damages muscles throughout the body, most notably the heart and skeletal muscle, and leads to general muscle weakness, breathing problems, and feeding difficulties. The onset and severity of disease symptoms vary widely based on the precise *GAA* variant(s) inherited [3]. The most severe phenotype, the classical infantile form, is clinically apparent in the first two months of life, causes cardiomyopathy, and is typically fatal in the first year of life [4]. The nonclassical infantile form of Pompe disease typically presents in the first year of life, progresses more slowly than the classical infantile onset form, and typically leads to respiratory failure without cardiomyopathy and death in later childhood. Later-onset forms of Pompe disease, with onset ranging from infantile to early adulthood, are also possible and progress at a variable rate [3].

Targeted enzyme-replacement therapy (ERT) for Pompe disease can effectively slow disease progression by providing sufficient enzyme to degrade glycogen [5]. Because ERT has the greatest benefit if started prior to the onset of symptoms [6] and individuals with Pompe disease are typically asymptomatic at birth, newborn screening for Pompe disease has been recommended by the Secretary of the U.S. Department of Health and Human Services [7] to identify children at risk for Pompe disease as early as possible.

The Missouri State Public Health Laboratory (MSPHL) initiated universal newborn screening for Pompe disease in January 2013, following guidance set by House Bill 716, the Brady Alan Cunningham Act [8]. This bill was signed into law in 2009 and mandated the expansion of newborn screening to include Pompe disease and several other lysosomal storage disorders (LSDs). This manuscript summarizes how MSPHL has refined its Pompe newborn screening protocols over the first six years in practice and highlights important findings through the follow-up program.

2. Materials and Methods

On 16 October 2012, MSPHL received a waiver from the Missouri Department of Health and Senior Services' Institutional Review Board, as implementation of LSD screening did not meet the IRB definition of research. MSPHL utilizes digital microfluidic fluorometry (DMF) (SEEKER; Baebies, Inc., Durham, NC) for Pompe disease screening [9–11]. Decreased enzyme activity of acid α-glucosidase (GAA) indicates an increased risk for Pompe disease. MSPHL initially established cutoffs based on the enzyme activity of diagnostic samples from patients with Pompe disease, which were provided by the Missouri genetic referral centers. These samples included newborn and nonnewborn samples from patients with infantile Pompe as well as later-onset Pompe. As more data was collected from routine newborn screening for Pompe, cutoffs were refined to better reflect newborn enzyme activity levels. As depicted in Figure 1, samples with measured enzyme activity above the instrument cutoff are presumed to be normal and no further testing is performed. For any samples with measured activity below the instrument cutoff, testing is repeated in duplicate and the average of the three values is calculated and used to assess risk. If the average value is below the referral cutoff, the baby is referred to a genetic referral center for evaluation and diagnostic confirmatory testing. If the average value is below the borderline cutoff but above the referral cutoff, an additional newborn screening sample is requested to repeat the dried blood spot (DBS) screening test. If the average value is above the borderline cutoff, the sample is presumed normal.

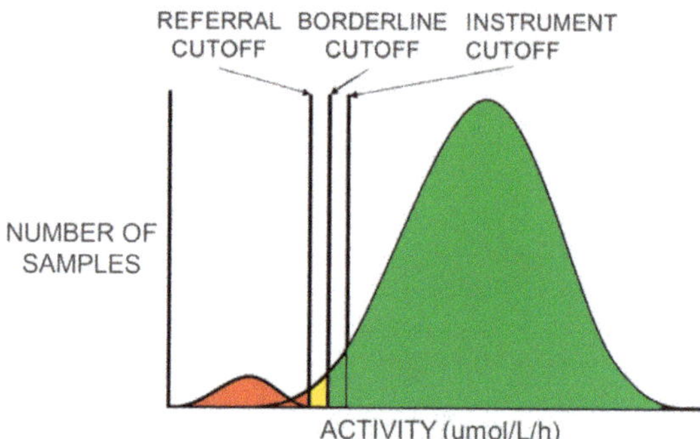

Figure 1. Representative depiction of the distribution of acid alpha-glucosidase (GAA) enzyme activity (low to high) in newborns. The small population to the left of the referral cutoff (depicted in red) indicates the high risk patient population that is referred for confirmatory testing. The large distribution to the right of the borderline cutoff (depicted in green) indicates the presumed normal population, who require no further action. As with many other newborn screening assays, there is an area of overlap between the affected and normal populations (indicated in yellow).

The cutoffs initially established by MSPHL did not account for age of the baby at sample collection, but after gathering population data over time, it was apparent that GAA enzyme activities decreased with the age of the newborn at the time of sample collection and that age-related cutoffs should be utilized. New cutoffs for samples from newborns >14 days of age at collection were implemented in mid-2013 and further refined as more data were collected and analyzed. MSPHL also discovered that lysosomal enzyme activities were affected by seasonal variability. Enzyme activities were reduced during the summer months, causing an increase in false positive referrals. After observing this trend, MSPHL implemented cutoff adjustments between winter and summer to reduce unnecessary referrals. Because four lysosomal enzymes are measured simultaneously with the DMF method, MSPHL leverages all four lysosomal enzyme results to ascertain sample quality. If multiple lysosomal enzyme activities are below their respective cutoffs, the sample is considered to have compromised quality and an additional newborn screening sample is requested. Whereas many other screening programs base GAA cutoffs on the percent of the daily mean or median, MSPHL has found that with the relatively small number of births per day in Missouri, fixed cutoffs that can be modified seasonally are more effective.

Because the need for second-tier testing for some LSDs was unrecognized when MSPHL began screening and there were no established second-tier testing options available at the time, MSPHL did not employ second-tier testing for Pompe disease. All presumptive positive screens were referred to a genetic referral center for diagnostic testing. In Missouri, there are four such centers to accept these patient referrals, and referrals are made based on predetermined geographic boundaries. The diagnostic testing performed for a presumptive positive patient is dependent on the referred condition; for Pompe disease, this testing may include leukocyte GAA enzyme activity, creatine kinase (CK) enzyme activity, urinary glucotetrasaccharide (HEX4), targeted gene sequencing, and cardiac evaluation. Additional testing may include quantification of lactic acid dehydrogenase (LDH), CK-MB (an isoenzyme of creatine kinase that is found mostly in the heart), liver enzymes (AST and ALT), and/or brain natriuretic peptide (BNP). While confirmatory testing for Pompe disease follows a general protocol, the procedure for each patient is highly dependent on the specific clinical presentation.

The predicted onset—infantile or late—is made during the time of confirmatory testing based on biochemical test results, imaging, clinical presentation including presence of cardiomyopathy, and variant analysis. Table 1 outlines the criteria used by the Missouri follow-up program to determine disease status. These are the general guidelines; the final classification is made by the follow-up team based on evaluation of all of the clinical information. Following a confirmation of infantile onset Pompe disease, treatment with enzyme replacement therapy is typically initiated as quickly as possible; in patients with a diagnosis of later-onset Pompe, treatment with ERT is typically delayed until the onset of symptoms or laboratory results consistent with progression of disease are observed.

Table 1. Missouri Newborn Screening Follow-Up Criteria for Pompe Disease.

Newborn Assessment	Classical Infantile	NonClassical Infantile	Later Onset	Genotype of Unknown Significance	Pseudodeficiency	Carrier
GAA enzyme activity	Absent or within affected range	Within affected range	Decreased	Decreased	Decreased	Decreased or normal
HEX4	Elevated	Elevated or WNL	WNL	WNL	WNL	WNL
Creatine Kinase (& other labs as indicated)	Elevated	Elevated or WNL	WNL	WNL	WNL	WNL
Chest x-ray, EKG, Echo	Abnormal	Mild abnormalities or WNL	WNL	WNL	WNL	WNL
Variant analysis	-Two pathogenic variants -One pathogenic variant and one or more VUS -Two VUS	-Two pathogenic variants -One pathogenic variant and one or more VUS -Two VUS	-Two pathogenic variants -One pathogenic variant and one or more VUS -Two VUS	-One infantile variant and one or more VUS -One late onset variant and one or more VUS -Two or more VUS	-Two pseudodeficiency alleles	-One pathogenic variant -May or may not be in combination with pseudo alleles
Clinical presentation	Muscle weakness, poor muscle tone, feeding issues, cardio-myopathy present	Muscle weakness or WNL	WNL at birth	WNL at birth	WNL at birth	WNL at birth

Abbreviations: WNL = within normal limits; VUS = variant(s) of unknown significance.

3. Results

3.1. Screening Results

In the first six years of LSD screening (January 2013 through December 2018), MSPHL tested approximately 467,000 newborns, of which 274 screened positive based on decreased GAA activity. Results of confirmatory testing for these specimens are presented in Table 2.

Table 2. Results of Confirmatory Pompe Testing.

Total Screened	~467,000
Screen Positives	274
Confirmed Disorders	46
Infantile Onset Pompe Disease	10
Later-onset Pompe Disease	36
Genotypes of Unknown Significance	8
Pseudodeficiencies	53
Carriers	65
Normal	97
Lost to Follow-up	5
Positive Predictive Value (PPV)	17.1%
False Positive Rate (FPR)	0.05%

Ten newborns were found to have infantile Pompe disease, and 36 were found to have later-onset Pompe disease. Eight newborns were found to have genotypes of unknown significance (GUS), and 51 newborns that screened positive were found to have pseudodeficiency variants. The false positive rate (FPR—0.05%) and the positive predictive value (PPV—17.1%) for the Pompe assay is comparable to published prospective screening results for GAA without the use of second-tier screening [12] and for other newborn screening tests in general [13].

Each genetic referral center was contacted to request follow-up information for newborns diagnosed with Pompe disease or a genotype of unknown significance. For each case, the referral center was asked to provide variant information and biochemical or other diagnostic test results. Additionally, referral centers were surveyed with specific questions about the current developmental status of each patient remaining in active follow-up. This survey collected binary responses (e.g., improved/unchanged/worsening) for symptoms in the following categories: cardiac, myopathy/hypotonia, respiratory, feeding, hearing, overall development, and laboratory test results (normal/abnormal).

Data from confirmatory testing for newborns confirmed with Pompe disease (infantile and later-onset) as well as newborns with genotypes of unknown significance was compiled (Table 3).

Table 3. Confirmatory Test Results for Patients with Pompe Disease or Genotypes of Unknown Significance (GUS).

Disease Classification	HEX4 (nmol/mol Creatinine)			Creatine Kinase (U/L)		
	n (Data Reported)	Median	Range	*n* (Data Reported)	Median	Range
Classical Infantile	7 (7)	22.7	13.4–38.6	7 (7)	662	466–3537
Nonclassical Infantile	3 (3)	5	3.7–25.2	3 (2)	416	398–435
Later-onset	36 (26)	4.65	2.3–12.3	36 (25)	127	50–466
GUS	8 (5)	6.6	2.3–7	8 (7)	87	71–203
Normal Range	<20 nmol/mol creatinine			<305 U/L		

3.2. Confirmed Positive Pompe Patients

3.2.1. Infantile Onset

Ten newborns were found to have infantile onset Pompe disease, with seven considered "classical" and three considered "nonclassical". Classical infantile Pompe is differentiated from nonclassical infantile Pompe by the presence of hypertrophic cardiomyopathy at birth. Of the classical infantile onset patients, all seven received testing for urine HEX4 and CK. Six (86%) had elevated HEX4 and seven (100%) had elevated CK activity. Six cases received a cardiac workup including EKG and echocardiogram; all six cases (100%) showed evidence of cardiomyopathy. The seventh case that did not receive a cardiac workup was diagnosed via amniocentesis, and had extremely elevated HEX4 and CK levels. All seven newborns began treatment with enzyme replacement therapy (ERT) at ages ranging from four days to one month. One of these patients was CRIM-negative and thus underwent immunosuppressive therapy prior to receiving ERT.

Three newborns were diagnosed with nonclassical infantile onset Pompe disease. Two of the three newborns were from the same family (separate births); both newborns were compound heterozygotes for the severe c.525DelT variant and the common later-onset c.-32-13T>G variant. These two newborns had normal HEX4 levels but mildly elevated CK. While the first newborn's HEX4 and cardiac workup were normal, the patient's liver enzymes (AST and ALT) were abnormal, CK-MB and LDH were elevated, and the newborn was failing to thrive. This newborn was initiated on enzyme replacement therapy at 29 days of age. The second newborn also had normal HEX4 but showed mild concentric left ventricular hypertrophy and elevated LDH. This newborn began treatment with ERT at 1 month of age. The third newborn with nonclassical infantile onset Pompe disease had significantly elevated HEX4 and CK. This newborn was compound heterozygous for the c.2560C>T variant, which is commonly associated with classical infantile onset Pompe disease and the c.2236T>C variant, which is a less common missense variant. This newborn began ERT at 1 month of age.

3.2.2. Later-Onset

Through screening, 36 newborns were identified and subsequently diagnosed with later-onset Pompe disease. Of these, 35 newborns received tests for HEX4 only ($n = 7$), CK only ($n = 9$), or both ($n = 19$). All newborns that received testing had normal results for HEX4 and six (21%) had mildly elevated CK. Ten of the newborns were found to have abnormal results for some combination of liver enzymes, LDH, CK-MB, and/or BNP.

Sixteen newborns received cardiac testing including a combination of chest x-ray, EKG, and echocardiogram—of which 15 were within normal limits. The other newborn in this group had an abnormal ECG with concern for right ventricular or biventricular hypertrophy. This newborn was compound heterozygous for the c.2560C>T variant and the c.-32-13T>G variant. During follow-up testing, this newborn exhibited worsening cardiomyopathy and hypotonia as well as abnormal labs; the newborn began ERT at 13 months of age.

3.2.3. Genotypes of Unknown Significance

Eight newborns were found to have genotypes of unknown significance. All eight received testing for either HEX4 only ($n = 1$), CK only ($n = 3$), or both ($n = 4$); all results were normal. Additionally, three of the newborns received an EKG and echocardiogram; these results were also normal. The genotypes of these eight patients were all different, and four of the eight had at least three detected variants. Two of these newborns remain in active follow-up and neither has begun to show clinical manifestations of Pompe disease. For two others, follow-up was deemed unnecessary unless clinical concerns arose, and the remaining four cases were lost to follow-up.

3.2.4. Pseudodeficiencies

Through screening, 53 newborns were found to have GAA pseudodeficiency. Three common GAA pseudodeficiency variants were represented in the Missouri population: c.1726G>A, c.2065G>A, and c.271G>A, including the common c.1726G>A/c.2065G>A haplotype. In total, the incidence of pseudodeficiency homozygotes or compound heterozygotes was 1:8811 (0.01%).

3.2.5. Other Results

Sixty five newborns were found through confirmatory testing to be heterozygous for a pathogenic variant, likely pathogenic variant, or variant of unknown significance. Additionally, 97 newborns were classified as normal due to normal confirmatory enzyme activities.

3.3. Current Follow-up Status

Of the 54 newborns identified with infantile onset or later-onset Pompe disease, or a genotype of unknown significance, 59% (32/54) remain in active follow-up with the genetic referral centers. This includes 7/10 infantile onset cases, 23/36 later-onset cases, and 2/8 with genotypes of unknown significance.

All ten patients with infantile onset Pompe disease—seven classical and three nonclassical—initiated enzyme replacement therapy at ages between four days and one month. Four of the classical infantile onset cases remain in active follow-up (two of the patients have moved out of the state and one is deceased). Cardiac symptoms (hypertrophic cardiomyopathy) have improved in all four active cases (4/4). In three of the cases, myopathy/hypotonia, growth, respiratory symptoms, hearing, feeding, and overall development have improved or remained unchanged since the initiation of ERT. In the remaining case, in which the patient has been receiving ERT for approximately 5.5 years, myopathy and hypotonia have worsened, hearing loss has occurred, and overall development has slowed.

All three of the newborns with nonclassical infantile onset remain in active follow-up. Cardiac symptoms, hypotonia, respiratory, hearing, feeding, and development status are unchanged in all three cases. In one case, ERT was discontinued within the first year of life due to infusion-related reactions. A desensitization protocol was attempted and unsuccessful. This patient is still followed closely and continues to have normal labs, normal growth, and no pulmonary concerns.

Of the 23 later-onset cases that are still in active follow-up, the status of 20/23 is unchanged and ERT has not been administered. In one case, cardiac symptoms improved without the aid of ERT; in another case, both cardiac symptoms and myopathy improved without ERT. In the final case, hypotonia worsened on follow-up and labs were abnormal, which resulted in the introduction of ERT at 13 months of age.

4. Discussion

4.1. Screening Results

Through nearly 6 years of prospective testing, Missouri screened approximately 467,000 newborns for GAA activity; as a result of this screening, 46 newborns were diagnosed with Pompe disease, including 10 with infantile onset Pompe disease. From the evidence report compiled in 2013 by the Condition Review Workgroup of the Advisory Committee on Heritable Diseases in Newborn and Children (ACHDNC), the incidence of Pompe disease in the United States was estimated at 1:28,000, with approximately 28% of the cases (1:100,000) presenting in the first 12 months of life (infantile onset) [4]. Missouri's screening program identified a higher than expected overall rate of Pompe disease (1:10,152) and infantile Pompe disease (1:46,700), with a slightly lower than expected percentage of cases diagnosed with infantile onset (22%).

The Missouri newborn screening laboratory overcame several challenges during the first months of screening for Pompe. As the first U.S. state to screen for Pompe, there was limited data available for the laboratory to reference when determining preliminary cutoffs, specifically for demographic variables. For example, the laboratory implemented age-related cutoffs for GAA activity after approximately four months of live screening when it was observed that median GAA activity decreases by more than 30% from birth to the 14th day of life. The decrease in median enzyme activity as a function of age at sample collection is illustrated in Figure 2. After implementation of age-related thresholds, the retest and false positive rates for the assay both decreased.

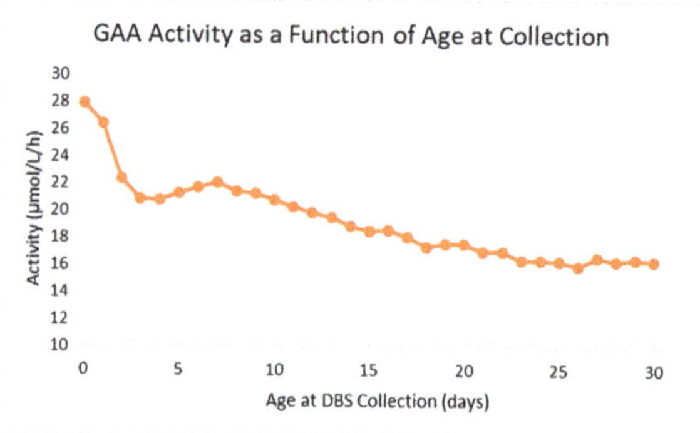

Figure 2. GAA enzyme activity decreases during the first 30 days of life. GAA enzyme activity data (Y axis) from this chart includes average screening results from 338,743 full-term newborns (>38 weeks gestational age) prospectively screened at MSPHL. Age at DBS sample collection (X axis) was rounded down to the nearest full day.

Similarly, the laboratory began to utilize the activity of all tested lysosomal enzymes as a measure of sample quality. This was beneficial with leukodepleted samples; since lysosomal enzymes in blood are predominantly present in white cells, activity of all lysosomal enzymes can be decreased in leukodepleted samples. Evaluation of all lysosomal enzymes is also useful during periods of high temperature and humidity. During the summer months, lysosomal enzymes may be denatured during extended exposure to elevated heat and humidity. Beginning the first summer following the start of screening, MSPHL has used all lysosomal enzyme activities as an indicator of sample quality and considers results that were abnormal for multiple lysosomal enzymes to be inconclusive, thereby requiring only a repeat/second newborn screen. The laboratory also adjusts the cutoffs during the summer months to account for the decrease in activity.

Comparison of HEX4 and CK, two diagnostic tests that are completed during the Pompe follow-up protocol, shows a correlation between disease severity and HEX4 and CK levels. The classical infantile onset patients all displayed elevated HEX4 and CK, while the nonclassical infantile and later-onset patients had results within the normal limits or mildly elevated results. Although CK activity for the nonclassical infantile cases was only mildly elevated, these cases were significantly elevated (median = 416) relative to the later-onset group (median = 126).

4.2. Variant Frequencies

4.2.1. Infantile and Later-Onset Variants

Of the 10 newborns diagnosed with infantile onset Pompe through screening, a total of 13 different variants were identified. Specific variant information is presented in Table 4. The c.525DelT variant was the most common infantile onset variant in this cohort, as was found in 3/10 cases, including two newborn family members prospectively identified through screening. The c.1447G>A, c.1802C>T, c.2560C>T, c.-32-13T>G, and exon 18 deletions were found in two patients each.

Table 4. Variants Identified in Infantile Pompe Disease Patients through Prospective Screening.

Diagnosis	Variants
Classical Infantile	c.1548G>A/del exon 18
Classical Infantile	c.1447G>A/c.2560C>T
Classical Infantile	c.670C>T/c.2481+31Del
Classical Infantile	c.525DelT/c.1447G>A
Classical Infantile	c.1827C>G/c.2662G>T
Classical Infantile	c.1802C>T/c.1802C>T
Classical Infantile	c.947A>G/del exon 18
Nonclassical Infantile	c.-32-13T>G/c.525DelT
Nonclassical Infantile	c.-32-13T>G/c.525DelT
Nonclassical Infantile	c.2560C>T/c.2236T>C

Of the later-onset patients, 36 different variants were detected. The most common variant detected was the c.-32-13T>G variant, which was found in 66.7% (24/36) patients, including 10 homozygotes. The c.841C>T variant was also found as part of a compound heterozygote in four different patients.

4.2.2. Pseudodeficiency

The frequency of pseudodeficiency variants was also evaluated. Of the 53 cases classified as pseudodeficiencies, variant information was available for 49 newborns. Of these 49 newborns, 48 (98%) had the common c.1726G>A/c.2065G>A pseudodeficiency haplotype. Two newborns also possessed the c.271G>A pseudodeficiency variant; one newborn was homozygous for this allele, and one newborn possessed this allele as a compound heterozygote with the c.1726G>A/c.2065G>A haplotype. When evaluating the ethnicities of the pseudodeficiency cases, 43 (88%) of the 49 cases with available variant information reported the newborn's ethnicity on the sample collection form as either "Asian", "Pacific Islander", or multiethnic including at least one of those two groups. These pseudodeficiency variants have previously been reported at high prevalence in the Asian population [14,15], and the results of screening in Missouri indicate that the vast majority of GAA pseudodeficiency cases are of Asian ethnicity. Based on the 2010 U.S. Census [16], 2.0% of Missouri's population is of Asian ethnicity, which is significantly below the U.S. average of 5.6%; other U.S. states or territories with a higher population proportion of Asian ethnicity may encounter higher rates of pseudodeficiency.

4.3. Current Follow-up Status

Fifty nine percent of newborns with Pompe disease or genotypes of unknown significance have maintained active follow-up after the initial diagnosis. All infantile onset patients remain in active follow-up with the exception of two families that no longer reside in the state; however, the proportion of later-onset patients and patients with genotypes of unknown significance that maintain active follow-up is far lower (23 of 36 later-onset, 64%, remain active; 2 of 8, 25%, GUS remain active).

Of the 10 infantile onset cases diagnosed, all 10 started enzyme replacement therapy. Nine of the infantile cases were CRIM-positive. Eight of ten have improved or unchanged symptoms, with one case of worsening hypotonia after more than 5 years on ERT, and another discontinuing ERT after less than one year following development of infusion-related reactions. Both of the newborns with worsening symptoms or adverse reactions were CRIM-positive. The CRIM-negative newborn has moved out of state and is no longer in active follow up in Missouri. One of the later-onset cases has developed symptoms consistent with Pompe disease and has initiated treatment; as this patient was 13 months of age at initiation of treatment, it reinforces that patients with later-onset Pompe disease may develop symptoms and require intervention with ERT in early childhood.

5. Conclusions

The Missouri newborn screening program continues to operate the longest continuous Pompe screening program in the United States and has screened approximately 467,000 newborns during its first six years of screening. Pompe screening has been very successful as 46 patients (approximately 1:10,000) have been diagnosed with Pompe disease, including 10 children (1:46,700) with infantile onset disorder. MSPHL also detected several common pseudodeficiency variants through screening, which caused an increased false positive rate. Implementation of second-tier testing would improve the FPR and PPV, as these pseudodeficiency variants are prevalent in the state's population; Missouri is currently evaluating options to implement second-tier testing for Pompe as well as use of postanalytical tools. Through analysis of screening results, the laboratory found that GAA activity decreases from birth to the 14th day of life and requires age-related cutoffs for the most appropriate risk assessment. MSPHL also found that testing for multiple lysosomal enzymes can aid in the determination of sample quality, as the lysosomal enzyme activities can be decreased in leukodepleted samples or in samples that are exposed to elevated heat and humidity.

To date, 11 patients (10 with infantile onset, 1 with later-onset) have begun treatment with enzyme replacement therapy through the screening and follow-up program. Early initiation of ERT has led to normal development and cardiac improvement for the majority of the infantile onset Pompe disease cases, and only one patient has discontinued ERT due to infusion-related reactions. Additionally, the presence of a patient with diagnosed later-onset disease that has already begun ERT offers a case study that later-onset disease may still present in infancy or early childhood.

Author Contributions: Conceptualization, J.L.K. and L.B.S.; formal analysis, T.L.K. and J.W.; investigation, T.L.K.; resources, J.L.K. and L.B.S.; data curation, L.B.S.; writing—original draft preparation, T.L.K. and J.W.; writing—review and editing, T.L.K., J.L.K., L.B.S., J.W., and C.B.; supervision, J.L.K. and T.L.K. All authors have read and agree to the published version of the manuscript.

Funding: This study received no external funding.

Acknowledgments: Special thanks to the Missouri Genetic Referral Centers (Cardinal Glennon Children's Hospital, Children's Mercy Hospital, St. Louis Children's Hospital, and University of Missouri Healthcare Children's Hospital) for providing follow-up and confirmatory data for these cases.

Conflicts of Interest: J.W. and C.B. have stocks or stock options in Baebies, Inc. All other authors declare no conflicts of interest.

References

1. Pompe, J.C. Over idiopathische hypertrophie van het hart. *Ned. Tijdschr. Geneeskd* **1932**, *76*, 304–311.

2. Bischoff, G. Zum klinischen Bild der Glykogen-Speicherungskrankheit (Glykogenose). *Z. Kinderheilkd.* **1932**, *52*, 722–726. [CrossRef]

3. Kishnani, P.S.; Beckemeyer, A.A.; Mendelsohn, N.J. The new era of Pompe disease: Advances in the detection, understanding of the phenotypic spectrum, pathophysiology, and management. *Am. J. Med. Genet. Part C Semin. Med. Genet.* **2012**, *160C*, 1–7. [CrossRef] [PubMed]

4. Kemper, A.R.; Comeau, A.M.; Prosser, L.A.; Green, N.S.; Tanksley, S.; Goldenberg, A.; Weinreich, S.; Ojodu, J.; Lam, M.K.K. Evidence Report: Newborn Screening for Pompe Disease by The Condition Review Workgroup: Chair Condition Review Workgroup. Available online: https://www.hrsa.gov/sites/default/files/hrsa/advisory-committees/heritable-disorders/rusp/previous-nominations/pompe-external-evidence-review-report-2013.pdf (accessed on 16 December 2019).

5. Kishnani, P.S.; Corzo, D.; Leslie, N.D.; Gruskin, D.; van Der Ploeg, A.; Clancy, J.P.; Parini, R.; Morin, G.; Beck, M.; Bauer, M.S.; et al. Early treatment with alglucosidase alfa prolongs long-term survival of infants with pompe disease. *Pediatr. Res.* **2009**, *66*, 329–335. [CrossRef] [PubMed]

6. Yang, C.-F.; Yang, C.C.; Liao, H.-C.; Huang, L.-Y.; Chiang, C.-C.; Ho, H.-C.; Lai, C.-J.; Chu, T.-H.; Yang, T.-F.; Hsu, T.-R.; et al. Very Early Treatment for Infantile-Onset Pompe Disease Contributes to Better Outcomes. *J. Pediatr.* **2016**, *169*, 174–180. [CrossRef] [PubMed]

7. Discretionary Advisory Committee on Heritable Disorders in Newborns DACHDNC Letter on the Addition of Pompe Disease to the Recommended Uniform Screening Panel. Available online: https://www.hrsa.gov/sites/default/files/hrsa/advisory-committees/heritable-disorders/reports-recommendations/letter-to-sec-pompe.pdf (accessed on 16 December 2019).

8. HB716|Missouri 2009|Establishes the Brady Alan Cunningham Newborn Screening Act Which Adds Certain Lysosomal Storage Diseases to the List of Required Newborn Screenings|TrackBill. Available online: https://trackbill.com/bill/missouri-house-bill-716-establishes-the-brady-alan-cunningham-newborn-screening-act-which-adds-certain-lysosomal-storage-diseases-to-the-list-of-required-newborn-screenings/40648/ (accessed on 16 December 2019).

9. Hopkins, P.V.; Campbell, C.; Klug, T.; Rogers, S.; Raburn-Miller, J.; Kiesling, J. Lysosomal storage disorder screening implementation: Findings from the first six months of full population pilot testing in Missouri. *J. Pediatr.* **2015**, *166*, 172–177. [CrossRef] [PubMed]

10. Hopkins, P.V.; Klug, T.; Vermette, L.; Raburn-Miller, J.; Kiesling, J.; Rogers, S. Incidence of 4 lysosomal storage disorders from 4 years of newborn screening. *JAMA Pediatr.* **2018**, *172*, 696–697. [CrossRef] [PubMed]

11. Sista, R.S.; Wang, T.; Wu, N.; Graham, C.; Eckhardt, A.; Winger, T.; Srinivasan, V.; Bali, D.; Millington, D.S.; Pamula, V.K. Multiplex newborn screening for Pompe, Fabry, Hunter, Gaucher, and Hurler diseases using a digital microfluidic platform. *Clin. Chim. Acta* **2013**, *424*, 12–18. [CrossRef] [PubMed]

12. Burton, B.K.; Charrow, J.; Hoganson, G.E.; Waggoner, D.; Tinkle, B.; Braddock, S.R.; Schneider, M.; Grange, D.K.; Nash, C.; Shryock, H.; et al. Newborn Screening for Lysosomal Storage Disorders in Illinois: The Initial 15-Month Experience. *J. Pediatr.* **2017**, *190*, 130–135. [CrossRef] [PubMed]

13. Kwon, C.; Farrell, P.M. The magnitude and challenge of false-positive newborn screening test results. *Arch. Pediatr. Adolesc. Med.* **2000**, *154*, 714–718. [CrossRef] [PubMed]

14. Chiang, S.C.; Hwu, W.L.; Lee, N.C.; Hsu, L.W.; Chien, Y.H. Algorithm for Pompe disease newborn screening: Results from the Taiwan screening program. *Mol. Genet. Metab.* **2012**, *106*, 281–286. [CrossRef] [PubMed]

15. Kumamoto, S.; Katafuchi, T.; Nakamura, K.; Endo, F.; Oda, E.; Okuyama, T.; Kroos, M.A.; Reuser, A.J.J.; Okumiya, T. High frequency of acid α-glucosidase pseudodeficiency complicates newborn screening for glycogen storage disease type II in the Japanese population. *Mol. Genet. Metab.* **2009**, *97*, 190–195. [CrossRef] [PubMed]

16. 2010 Census Data|Office of Administration. Available online: https://oa.mo.gov/budget-planning/demographic-information/2010-census-data (accessed on 16 December 2019).

 International Journal of
Neonatal Screening

Article

Expanding Newborn Screening for Pompe Disease in the United States: The NewSTEPs New Disorders Implementation Project, a Resource for New Disorder Implementation

Kshea Hale [1,*], Yvonne Kellar-Guenther [2], Sarah McKasson [3], Sikha Singh [1] and Jelili Ojodu [1]

[1] Association of Public Health Laboratories, Silver Spring, MD 20910, USA; sikha.singh@aphl.org (S.S.); jelili.ojodu@aphl.org (J.O.)

[2] Center for Public Health Innovation, Littleton, CO 80120, USA; ykellar-guenther@ciinternational.com

[3] Colorado School of Public Health, Department of Epidemiology, University of Colorado, Anschutz Medical Campus, Aurora, CO 80045, USA; sarah.mckasson@cuanschutz.edu

* Correspondence: kshea.hale@aphl.org; Tel.: +1-240-485-3842

Received: 12 May 2020; Accepted: 8 June 2020; Published: 11 June 2020

Abstract: Public health programs in the United States screen more than four million babies each year for at least 30 genetic disorders. The Health and Human Services (HHS) Advisory Committee on Heritable Disorders in Newborns and Children (ACHDNC) recommends the disorders for state newborn screening (NBS) programs to screen. ACHDNC updated the Recommended Uniform Screening Panel (RUSP) to include Pompe disease in March 2015. To support the expansion of screening for Pompe disease, the Association of Public Health Laboratories (APHL) proposed the Newborn Screening Technical assistance and Evaluation Program (NewSTEPs) New Disorders Implementation Project, funded by the HHS' Health Resources and Services Administration (HRSA) Maternal and Child Health Bureau (MCHB). Through this project, APHL provided financial support to 15 state NBS programs to enable full implementation of screening for Pompe disease. As of 27 April 2020, nine of the 15 programs had fully implemented Pompe disease newborn screening and six programs are currently pursuing implementation. This article will discuss how states advanced to statewide implementation of screening for Pompe disease, the challenges associated with implementing screening for this condition, the lessons learned during the project, and recommendations for implementing screening for Pompe disease.

Keywords: newborn screening; Pompe disease; new disorders implementation

---

1. Introduction

Public health programs in the United States screen more than four million babies each year for at least 30 genetic disorders. Through the newborn screening (NBS) program, states identify newborns at risk for certain metabolic, endocrine, hemoglobin, neuromuscular, and other inherited conditions. To guide this process, the Health and Human Services (HHS) Advisory Committee on Heritable Disorders in Newborns and Children (ACHDNC) recommends the disorders for state NBS programs to screen [1]. Each disorder is selected through a vigorous evidence review process. Once approved, these disorders are included on the Recommended Uniform Screening Panel (RUSP).

There are 35 core disorders on the RUSP, as of April 2020 [2]. ACHDNC updated the RUSP to include Pompe disease in March 2015.

Pompe disease is a multi-systemic lysosomal storage disorder (LSD) caused by a deficiency in the acid alpha glucosidase (GAA) enzyme. This enzyme helps lysosomes breakdown glycogen, a complex sugar, in the body. Without this enzyme, glycogen accumulates in the cells, causing them to lose their ability to function properly. If left untreated, babies born with the most severe form of Pompe disease die within their first year of life [3]. In the United States, Pompe disease affects approximately one in 40,000 births. Implementing NBS for Pompe disease allows babies with the disorder to be identified and linked to care, often before symptoms arise. As of April 2020, 23 states in the US screen for Pompe disease, nine of which received direct funding and technical assistance from the Association of Public Health Laboratories (APHL) to implement screening for Pompe disease.

To implement universal screening for Pompe disease, there are many factors for state NBS programs to consider. This includes determining the cost to screen for the disorder, gaining legislative and funding approval to screen, securing fee increases if necessary, selecting a technology and methodology to screen, expanding laboratory space and capacity, acquiring the appropriate equipment, hiring laboratory and follow-up staff, and developing and implementing education initiatives for the public and for clinical providers. While these factors are important for the addition of all new disorders to state screening panels, Pompe disease introduced unique considerations for the newborn screening system. States screening for this disorder must consider establishing algorithms for the detection of infantile Pompe disease in distinction from late onset Pompe disease, defining the duration of short term follow up and late onset Pompe disease and defining the role of NBS in long term follow up (LTFU) for individuals identified with the disorder.

2. Methods

APHL was awarded a cooperative agreement from the HHS' Health Resources and Services Administration (HRSA) Maternal and Child Health Bureau (MCHB) to expand the ability of state NBS programs to screen for Pompe disease. Through the New Disorders Implementation Project, the APHL Newborn Screening Technical assistance and Evaluation Program (NewSTEPs) provided financial support over the course of three years to 15 state NBS programs to enable and accelerate full implementation of newborn screening for Pompe disease [4].

APHL also selected and funded three additional states to serve as Peer Network Resources Centers (PNRCs) during the course of the New Disorders Implementation Project. The PNRCs served as content area experts providing technical assistance and training regarding laboratory techniques, follow-up procedures, and educational processes related to Pompe disease, as well as for other new disorders screening as needed. Each state NBS program and PNRC applied to participate in the project through a competitive request for proposals (RFP) and a subsequent evaluation process. In the RFP, NewSTEPs proposed a four-phase model to stratify the NBS implementation process, as seen in Figure 1 [5]. In their RFP response, each state program indicated where they were in the implementation of screening for Pompe disease, which other new disorders they were seeking support for, what they hoped to accomplish with this funding opportunity, and how they would use the funding.

The 15 participating state NBS programs provided a combination of monthly and annual progress reports where they described activities and progress toward full implementation and also participated in monthly calls with NewSTEPs staff to track and describe their activities and indicate their successes, challenges, needs, lessons learned, and recommendations.

Phase 1- Legislative/Mandate Stage

Newborn Screening Programs that require assistance/guidance for adding new conditions to the required list in the state, including but not limited to fee increases to support capital and ongoing expenditures and legislative mandates.

Phase 2- Logistics/Testing Implementation and Development of Follow-Up Network Stage

Programs that require assistance obtaining equipment and training to perform new condition screening.

Phase 3- Education/Information Dissemination Stage

Programs that require assistance developing and fully implementing new condition NBS education initiatives.

Phase 4- Full Implementation Stage

New condition NBS is required and offered to all newborns and the new condition education materials are appropriate for all audiences.

Figure 1. Phases of Implementation for New Disorders Newborn Screening.

3. Results

Of the 15 funded state newborn screening programs, nine were also funded to implement screening for Mucopolysaccharidosis type I (MPS I) and X-linked adrenoleukodystrophy (X-ALD) simultaneously with Pompe disease and six were funded to implement MPS I simultaneously with Pompe disease.

Additionally, the 15 funded programs set varied project goals and objectives to move them toward full implementation of screening for Pompe disease during the funding period. A summary of these goals are found in Table 1.

Table 1. Goals being pursued by the 15 Association of Public Health Laboratories (APHL) funded states for Pompe disease newborn screening implementation.

Project Goals for Pompe Disease Implementation	Number of States (% of Total, n = 15)
Full implementation of population screening	5 (33%)
Implementation of education or follow-up activities	4 (27%)
Validation activities and pilot screening	6 (40%)

3.1. Progress toward Pompe Disease NBS Implementation

Upon initiation of the funded project, two of the 15 state NBS programs were working on Phase 1 (Figure 1) to obtain approval and secure fee increases from their NBS Advisory Committee, Board of Health Commissioner, and/or State Budget Authority to screen for Pompe disease.

Upon initiation of the funding project, nine state NBS programs reported that they were working on Phase 2 (Figure 1) toward ensuring infrastructure and staffing readiness for Pompe disease newborn screening. Activities during this time included identifying screening methodologies, acquiring equipment, developing algorithms and protocols for screening, determining cut-off values for analysis, establishing contracts with medical specialists and treatment centers, developing long-term follow-up programs, pursuing assay validation, and integrating Pompe disease reporting features into their Laboratory Information Management Systems (LIMS).

Upon initiation of funding, three state NBS programs reported that they would be working toward Phase 2 (Figure 1) in securing infrastructure and staffing readiness simultaneously with Phase 3 (Figure 1) in developing educational materials for providers, families, and the general public as well as implementing educational and outreach activities.

Upon initiation of funding, only one state NBS program reported that they would be engaging in Phase 1, Phase 2, and Phase 3 (Figure 1) simultaneously.

In their applications, 10 state NBS programs anticipated that they would have implemented Pompe disease screening at the end of the project period, one program anticipated that they would have entered Phase 3 (Figure 1), and two programs anticipated that they would have entered Phase 2 (Figure 1) at the conclusion of the funding period.

3.2. How States Advanced to Statewide Implementation of Screening for Pompe Disease

Participating state NBS programs utilized funding from the NewSTEPs New Disorder Implementation Project to purchase equipment and supplies, develop and translate educational materials, develop a statewide outreach program, host a community engagement event, hire laboratory and follow-up personnel, purchase software for sequence analysis, travel to professional meetings or to PNRCs for training, configure LSDs workflow in Laboratory Information Management Systems (LIMS), and develop a long-term follow-up program (Table 2).

Table 2. Utilization of funding by states to implement screening for Pompe disease.

Project Expenses	Number of States ($n = 15$)
Equipment and supplies	2
Staffing	1
Educational activities	0
Travel	0
Other expenses	0
Equipment, supplies, and travel	2
Equipment, supplies, and educational activities	1
Staffing, travel, educational activities, and other expenses	2
Equipment, supplies, staffing, educational activities, and other expenses	2
Equipment, supplies, staffing, travel, educational activities, and other expenses	3
Equipment, supplies, travel, educational activities, and other expenses	1
Equipment, supplies, staffing, and other expenses	1

Each program engaged at least one PNRC for technical assistance during the project. The PNRCs provided states with LSD standard operating procedures and algorithms, sample panels for validation, technical assistance around LSD test validation, LSD cost data, comparison analysis, second-tier testing guidance for Pompe disease, DNA sequence analyses for Pompe disease, and Pompe disease related educational materials for primary care providers and parents. The PNRCs also retested abnormal LSDs for state programs. One PNRC hosted an annual Tandem Mass Spectrometry, DNA Sequencing, and a Follow-up hands-on workshop. Another program prepared Pompe disease follow-up flow diagrams. State NBS programs also visited PNRCs for direct one-on-one-training to learn laboratory and follow up best practices and to observe processes.

As of 27 April, 2020, nine of the 15 programs that participated in the project had fully implemented Pompe disease newborn screening, offering the screening universally to all newborns in the following states: California, Michigan, Minnesota, Nebraska, New Jersey, Ohio, Tennessee, Virginia, and Washington. The other six programs are currently pursuing screening implementation for Pompe disease.

Programs participating in the NewSTEPs New Disorders Implementation project to implement Pompe Disease Newborn Screening also achieved the following milestones during the course of the funding period: created and deployed an online education module for providers and parents, educated 500 health care professional across the state regarding Pompe disease, implemented second-tier testing for Pompe disease, developed a follow-up quality improvement plan, and hosted a successful deliberative community engagement event.

It is important to note that while the activities for the 15 state newborn screening programs highlighted here are specific to their pursuit of Pompe disease newborn screening, additional disorders (MPS I and x-ALD) were also pursued by a number of these programs. There are several aspects of the newborn screening multiplexing of the screening assay, development of educational materials, and training of implementation activities described above that are generalizable to these other disorders, including laboratory and follow-up staff, establishment of reporting algorithms, among others.

4. Challenges

The participating state NBS programs faced several challenges implementing screening for Pompe disease. As noted above, while many barriers and challenges in implementing newborn screening are generalizable to multiple disorders, the challenges presented below are articulated by the 15 newborn screening programs pursuing Pompe disease newborn screening during the course of the NewSTEPs New Disorders Implementation Project. These challenges are as follows (Table 3):

Table 3. Challenges reported by participating state newborn screening programs when implementing Pompe Disease Newborn Screening during the course of the NewSTEPs New Disorder Implementation Project.

Challenge	Reason Provided
Hiring	Lengthy in-state human resources approval processes Lack of qualified applicants to perform enzyme activity detection and/or genetic testing Lack of qualified applicants to analyze results and perform risk assessments for the various stages of Pompe disease onset
Retention of Staff	Availability of more lucrative positions elsewhere Staff retirements coupled with lack of backfilling these vacancies
Procurement Delays	Complicated and time-consuming state procurement processes Delays by the state procurement office in establishing contracts with vendors to purchase reagents and lease instrumentation to perform the assay for Pompe disease
Instrumentation Challenges	New instrumentation did not perform as expected
Infrastructure Challenges	Delays in the construction of adequate laboratory space to house instrumentation required for the addition of Pompe disease newborn screening
Information Technology	Delays by vendors to complete upgrades to Laboratory Information Management Systems (LIMS) to accommodate results reporting and messaging for newly added disorder (Pompe disease) to the workflow Time and expense involved in gathering system requirements and working with Information Technology (IT) developers and informatics support to modify LIMS and vendor specific software
Clinical Follow-up	The pediatric community and other healthcare providers who would be responsible for caring for babies with an abnormal result have limited knowledge regarding Pompe disease

5. Lessons Learned

When implementing screening for Pompe disease the participating state newborn screening programs recommended allowing for sufficient time to accomplish procurement of instrumentation required for first tier detection of GAA enzyme activity. Multiple methodologies exist to perform a Pompe disease newborn screening assay, including the use of digital microfluidics or tandem mass spectrometry, and programs recommended building in sufficient time to identify the methodology to use [6]. In addition, delays should be anticipated when performing building and space modifications to accommodate Pompe disease newborn screening into the existing workflow. In order to address these delays associated with introducing Pompe disease newborn screening, while maintaining existing quality practices, adequate staffing should be available.

Before implementing full population screening for Pompe disease, the participating states noted that NBS programs should determine second-tier needs, address special populations since these infants can potentially influence the determination of cut-off values for assay analysis and may require repeat screening, and ensure LIMS integration is complete. The introduction of Pompe disease newborn screening can increase the workload for all staff involved in the testing and reporting process, therefore affecting communication with providers.

6. Recommendations

At the end of the NewSTEPs New Disorder Implementation Project, the participating state NBS programs provided recommendations for implementing newborn screening for Pompe disease. The recommendations included participating in APHL NewSTEPs committees and in-person meetings to collaborate and connect with other programs and NBS stakeholders, creating standardized educational materials and long-term follow-up strategies that can be modified for state-specific use, and ensuring that all state stakeholders including pediatric subspecialties and parent representatives are involved in discussing all plans for implementation. Other recommendations include adding second-tier screening to reduce false positives and monitoring screening outcomes and refining cut-off reporting algorithms. Prior to implementing second-tier screening or refining cut-offs, states' programs screening for Pompe disease using tandem mass spectrometry or digital microfluidics reported higher than expected false positive and referral rates, increasing the patient load at NBS follow-up coordinating and genetic centers. Implementing second-tier testing and refining cut-offs reduced false positive and referral rates. Participating state NBS programs also recommended developing long-term follow-up guidelines for cases of Late Onset Pompe Disease (LOPD).

7. Conclusions

The ACHDNC updated the RUSP to include Pompe disease in March 2015. To support the expansion of newborn screening across the United States for Pompe disease, APHL proposed the NewSTEPs New Disorders Implementation Project, funded by HRSA. Through this project, 15 state NBS programs received direct funding through a competitive RFP process to implement screening for Pompe disease and all states had access to technical assistance and training via PNRCs. To provide further support to state NBS programs with the implementation of Pompe disease, APHL hosted national meetings to provide a platform for state programs to share their experiences implementing screening, hosted disease-specific webinars, and convened a NewSTEPs New Disorders Workgroup to serve as educational experts. As a result of the project, all states had access to information and resources regarding Pompe disease screening implementation.

To ensure that all states have the capacity to screen for Pompe disease and newborns living with this disorder are identified and linked to care, APHL will continue to provide support through technical assistance, training, webinars, quality improvement initiatives, and national meetings.

Int. J. Neonatal Screen. **2020**, *6*, 48

Author Contributions: Conceptualization, S.S., Y.K.-G., and J.O.; methodology, K.H. and Y.K.-G.; data curation, K.H.; writing—original draft preparation, K.H. and S.S.; writing—review and editing, K.H., S.S., Y.K.-G., J.O., and S.M.; visualization, K.H. and S.M.; supervision, S.S.; project administration, K.H.; funding acquisition, J.O. All authors have read and agreed to the published version of the manuscript.

Funding: This project was supported by the Health Resources and Services Administration (HRSA) of the U.S. Department of Health and Human Services (HHS) under grant number # UG9MC30369 New Disorders Implementation Project for $4,000,000. This information or content and conclusions are those of the author and should not be construed as the official position or policy of, nor should any endorsements be inferred by HRSA, HHS or the U.S. Government.

Acknowledgments: The information in this article is based on participation from NBS programs in the following states/territories: California, Florida, Georgia, Iowa, Michigan, Minnesota, Missouri, Nebraska, New Jersey, New York, North Carolina, Ohio, South Carolina, Tennessee, Utah, Virginia, Washington, and Wisconsin.

Conflicts of Interest: The authors declare no conflict of interest. The funders had no role in the design of the study; in the collection, analyses, or interpretation of the data; in the writing of the manuscript, or in the decision to publish the results.

References

1. Health Resources and Services Administration. Advisory Committee on Heritable Disorders in Newborns and Children. Available online: https://www.hrsa.gov/advisory-committees/heritable-disorders/index.html (accessed on 7 May 2020).

2. Health Resources and Services Administration. Recommended Uniform Screening Panel. Available online: https://www.hrsa.gov/advisory-committees/heritable-disorders/rusp/index.html (accessed on 7 May 2020).

3. National Institutes of Health; U.S. National Library of Medicine. Pompe Disease. Available online: https://ghr.nlm.nih.gov/condition/pompe-disease (accessed on 26 November 2019).

4. Ojodu, J.; Singh, S.; Kellar-Guenther, Y.; Yusuf, C.; Jones, E.; Wood, T.; Baker, M.; Sontag, M.K. NewSTEPs: The Establishment of a National Newborn Screening Technical Assistance Resource Center. *Int. J. Neonatal Screen.* **2018**, *4*, 1. [CrossRef]

5. APHL. Request for Proposals (RFP): Support for Newborn Screening Implementation of New RUSP Disorders. Available online: https://www.aphl.org/rfp/Documents/2017_RFP_Y02%20_New_Disorders_Project_Aug%202017.pdf (accessed on 7 May 2020).

6. Millington, D.S.; Bali, D.S. Current State of the Art of Newborn Screening for Lysosomal Storage Disorders. *Int. J. Neonatal Screen.* **2018**, *4*, 24. [CrossRef]

International Journal of
Neonatal Screening

Article

At-Risk Testing for Pompe Disease Using Dried Blood Spots: Lessons Learned for Newborn Screening

Zoltan Lukacs [1], Petra Oliva [2,*], Paulina Nieves Cobos [1], Jacob Scott [2], Thomas P. Mechtler [2] and David C. Kasper [2]

[1]   Newborn Screening and Metabolic Diagnostics Unit, Hamburg University Medical Center, 20251 Hamburg, Germany; lukacs@uke.de (Z.L.); acquim@hotmail.com (P.N.C.)
[2]   ARCHIMED Life Science GmbH, 1110 Vienna, Austria; j.scott@archimedlife.com (J.S.); t.mechtler@archimedlife.com (T.P.M.); d.kasper@archimedlife.com (D.C.K.)
*   Correspondence: p.oliva@archimedlife.com

Received: 31 January 2020; Accepted: 9 December 2020; Published: 21 December 2020

Abstract: Pompe disease (GSD II) is an autosomal recessive disorder caused by deficiency of the lysosomal enzyme acid-α-glucosidase (GAA, EC 3.2.1.20), leading to generalized accumulation of lysosomal glycogen especially in the heart, skeletal, and smooth muscle, and the nervous system. It is generally classified based on the age of onset as infantile (IOPD) presenting during the first year of life, and late onset (LOPD) when it presents afterwards. In our study, a cohort of 13,627 samples were tested between January 2017 and December 2018 for acid-α-glucosidase (GAA, EC 3.2.1.20) deficiency either by fluorometry or tandem mass spectrometry (MS). Testing was performed for patients who displayed conditions of unknown etiology, e.g., CK elevations or cardiomyopathy, in the case of infantile patients. On average 8% of samples showed activity below the reference range and were further assessed by another enzyme activity measurement or molecular genetic analysis. Pre-analytical conditions, like proper drying, greatly affect enzyme activity, and should be assessed with measurement of reference enzyme(s). In conclusion, at-risk testing can provide a good first step for the future introduction of newborn screening for Pompe disease. It yields immediate benefits for the patients regarding the availability and timeliness of the diagnosis. In addition, the laboratory can introduce the required methodology and gain insights in the evaluation of results in a lower throughput environment. Finally, awareness of such a rare condition is increased tremendously among local physicians which can aid in the introduction newborn screening.

Keywords: newborn screening; Pompe disease; dried blood spots; Pompe disease diagnostics testing

---

1. Introduction

Pompe disease (GSD II) is an autosomal recessive disorder caused by deficiency of the lysosomal enzyme acid-α-glucosidase (GAA, EC 3.2.1.20), leading to a generalized accumulation of lysosomal glycogen especially in the heart, skeletal and smooth muscle, and the nervous system. It is generally classified based on the age of onset as infantile (IOPD) presenting during the first year of life, and late onset (LOPD) when it presents afterwards [1–3]. IOPD is usually associated with cardiomyopathy and then referred to as classic Pompe disease [1]. Infants with classic Pompe disease typically present during the first few weeks of life with hypotonia, progressive weakness, macroglossia, and hepatomegaly. Most of these infants die by their first birthday [4]. In the rare instances presenting without cardiomyopathy, it is referred to as non-classic Pompe disease [5–7].

The diagnosis of Pompe disease is usually made based on the typical clinical presentation followed by the demonstration of deficiency of GAA enzyme activity in muscle, skin fibroblasts or more recently dried blood spots (DBS) as well as GAA mutation analysis [2,8]. Diagnosis of Pompe disease through newborn screening (NBS) is also possible [9,10]. Prior to the initiation of enzyme replacement therapy, rapid determination of CRIM status [11,12] in patients with infantile onset Pompe disease is extremely important [13]. Depending on results immune-suppressive therapy may be initiated. Pompe disease is still considered to be a rare inborn error of metabolism with an estimated frequency of about 1/40,000 and a higher incidence in certain populations such as African Americans (1/14,000), Northern Europeans of Dutch origin and South East Asians [2]. Early results of newborn screening pilot studies from Taiwan [14] and the USA [15–17] indicated a higher general incidence, especially of LOPD cases which may be missed as clinical symptoms are less clear.

GAA catalyzes the hydrolysis of $\alpha1\rightarrow4$ glucosidic linkages in glycogen at acid pH [2]. Specificity for the natural substrate (glycogen) is gained during its maturation. The activity of mature (76/70-kDa) GAA for its natural (glycogen) substrate is considerably more robust than its activity towards the artificial substrate (4-methylumbelliferyl-α-D-glucoside; 4-MU), which is frequently used in in-vitro assays. However, 4-MU is also a substrate for several other enzymes including "leukocyte" neutral isoenzymes, glucosidase II (GANAB), neutral α-glucosidase C (GANC), and maltase glucoamylase (MGAM). Therefore, using maltose or, preferably, acarbose as an inhibitor of MGAM activity, allows for the measurement of GAA activity in DBS samples with minimal interference by other α-glucosidases. This assay serves as the basis for newborn screening and the non-invasive diagnosis of Pompe disease [18–20]. As a result, multiplex newborn screening assays for Pompe disease (based on GAA enzyme activity) and other lysosomal storage disorders using fluorometry, digital microfluidics or tandem mass spectrometry have been developed [10,21–23]. In addition for qualitative and quantitative assessments of the disease burden, and clinical measurement of the impact of Pompe disease on various affected systems, urinary glucose tetrasaccharide (Glc4), a biomarker of glycogen storage with 94% sensitivity and 84% specificity for Pompe disease, is frequently used in monitoring the response of patients to enzyme replacement therapy and as an adjunct to acid α-glucosidase activity measurements [24–26]. However, there is still no reliable biomarker to predict the natural course of the disease in an individual or the point of time when ERT should be administered for LOPD cases. This complicates the introduction of newborn screening in many areas. In contrast, diagnostic testing allows early detection of LOPD cases without ethical problems and may pave the way for a future introduction of whole population Pompe screening.

In this paper, we present the data from testing over 13,000 individuals suspected having of Pompe disease collected over a two-year period by two different centers in Europe (Hamburg, Germany and Vienna, Austria). At-risk testing is the use of the assay to determine whether an individual at increased risk of having Pompe disease (because they have family history or symptoms remotely associated with the disease but not pathognomic for the disease such as CK-elevations of unknown origin) has a deficiency of GAA. Further diagnostic testing will be required for individuals whose test is suggestive of the condition to confirm true deficiency of enzyme or for example a compromised sample, as sample quality and shipping conditions can affect the results.

2. Materials and Methods

A DBS kit containing a customized card (Whatman 903) for blood sampling and sampling instructions, and an envelope was provided to physicians upon request for α-glucosidase testing. Dried blood specimens were received with brief clinical details and an ICF (informed consent) between January 2017 through December 2018. DBS protocols used to measure α-glucosidase (EC:3.2.1.20) enzyme activities are given below.

2.1. Fluorometric Method

The method by Chamoles et al. [18] was slightly modified. A 3 mm DBS was eluted with 360 µL of demineralized water, then 40 µL aliquots transferred to a 96-well plate. The test was run using the artificial substrate 4-methylumbelliferyl-α-D-glucoside (1.4 mM, Sigma-Aldrich, Darmstadt, Germany) in 40 mM sodium acetate (CH3COONa) buffer at pH 3.8 (Merck, Darmstadt, Germany) with and without the addition of 10 µL of 80 µM acarbose solution (Toronto Research Chemicals, North York, ON, Canada).

The assay was also performed at pH 7.0 (40 mM CH3COONa buffer adjusted with hydrochloric acid/sodium hydroxide (HCl/NaOH) to pH 7.0), to assess the quality of the DBS. The α-glucosidase activity at pH 7.0 is a convenient tool for quality assessment because the same buffers adjusted to pH 7.0 can be used as for the target enzyme at pH 3.8. In addition, the enzyme activity at pH 7.0 is usually less stable than α-glucosidase activity at pH 3.8 when subjected to detrimental pre-analytical conditions, thereby being an early marker for specimen quality. All tests were run in duplicate. After incubation for 21 h at 37 °C, the reaction was stopped by the addition of 200 μL EDTA buffer (150 μM, pH 11.5; Sigma-Aldrich). The 40 μL of DBS eluate that had been stored at 4 °C overnight was added to specific wells that served as blanks. A standard curve of 4-methylumbelliferone (0 to 3 μM (Sigma-Aldrich)) run on each plate was used for the calculation of enzyme activities. The fluorescence was read on a Victor D instrument (Wallac Oy, Turku, Finland) or a BioTek Synergy H1 (Bad Friedrichshall, Germany). In addition to enzyme activity, the percent inhibition with acarbose and the ratio of α-glucosidase activities at pH 3.8 with inhibition to the activity at pH 7.0 were calculated to aid in the interpretation of results [27]. For samples from patients older than 1 year of age, a truncated assay which relied only on the measurement of α-glucosidase with acarbose inhibition was used. As a reference enzyme, β-galactosidase was run on these samples. Specimens from infants and those that showed diminished α-glucosidase activity in the truncated test, were analyzed using the test with and without acarbose. The calculation of additional ratios allowed for a better interpretation of results from samples with borderline values. Individuals from whom specimens with normal results in the truncated assay were considered not affected by Pompe disease and this assay allowed higher throughput testing and expedited reporting of results.

2.2. MS Method

The assay was based on previously published methods [9,28]. The samples were processed using the following steps. The activities of acid β-glucocerebrosidase (ABG; Gaucher disease), acid sphingomyelinase (ASM; Niemann-Pick A/B disease), α-glucosidase (GAA; Pompe disease) and α-galactosidase; GLA; Fabry disease) were measured in a multiplex assay. The extract from one 3.2 mm punch per DBS sample was combined with substrate and internal standard (S&IS) mixtures. Incubation was performed at 37 °C for 20–22 h. The reaction was stopped by adding 100 μL stopping solution (80% acetonitrile plus 0.2% formic acid in water). Aliquots were transferred to a new deep-well plate, covered with aluminum foil and centrifuged at 3000× *g* for 15 min prior to mass spectrometry analysis. Background activity of a blank blood collection paper spot was subtracted from the DBS activity. Two QC samples with previously established activity levels for each enzyme and heat inactivated samples were included in each plate as assay controls.

3. Results

Between January 2017 and December 2018 13,627 specimens were tested for GAA deficiency by fluorometry or tandem mass spectrometry (MS) using Dried Blood Spots (DBS) (Table 1). Specimens came from 51 different countries but most were from Germany, Poland, Turkey, Italy and Iran. Approximately 30% of all samples submitted were from infants. The median age was 17 years and the range 1–95 years. Specimens from individuals with family history of the condition were excluded.

Table 1. Number of samples tested between January 2017 and December 2018 at the specialized centers in Hamburg, Germany and Vienna, Austria.

	Fluorometry Method	MS Method
Number of tests	7340	6287
Normal enzyme activity	6921	5591
Enzyme activity below cut-off (positive)	419 (6%)	696 (11%)

Most of the tested samples (92%) showed normal enzyme activity. 8% of the samples showed enzyme activity below the respective cut-off. In most cases with decreased enzyme activity (419 from the fluorometric method and 696 from the MS method) genetic analysis was performed on the same bloodspot. 35–40% of the low enzyme samples screened by the MS method were genetically confirmed with two pathogenic variants, and similar confirmation rates have been obtained for the fluorometric method. Fluorometry has lower sensitivity in comparison to mass spectrometry [29].

Some positive results were obtained in specimens affected by detrimental pre-analytical conditions which did not necessarily reduce other enzyme activities. In those instances, a second card was requested for analysis.

Unfortunately, not all DBS came with a description of clinical symptoms. However, the major symptoms that were given are summarized in Table 2 grouped by analogous symptoms and sorted by severity. Cardiomyopathy was present almost exclusively in infant cases while CK-elevations of unknown origin or limb girdle muscle dystrophy of unknown origin were more prominent among LOPD cases in the Hamburg cohort.

Table 2. Clinical symptoms provided for samples submitted to the study centers in Hamburg and Vienna. Not all specimens contained such information.

	Clinical Presentation
1	Cardiomyopathy
2	Hypotonia—floppy baby, proximal and progressive muscle weakness, limb girdle muscle weakness, muscle pain, loss of strength and myalgia
3	Scoliosis myopathy, rigid spine, diffuse myopathy, myopathic syndrome, EMG: myogenic involvement, motor deficit of the belt
4	Elevated biomarkers—CK, myoglobin, transaminases

Statistical results of the GAA enzyme activity measurement using either fluorimetry or tandem mass spectrometry are listed in Table 3.

Table 3. Statistical results for both reference methods used in Hamburg (fluorometry) and Vienna (MS). For the normal values all specimens from individuals considered unaffected by Pompe disease have been included.

	Fluorometry Method [μmol/punch/h]		Tandem Mass Spectrometry Method [μmol/L/h] with Acarbose
	α-Glucosidase with Acarbose	α-Glucosidase without Acarbose	
Mean	7.43×10^{-8}	1.24×10^{-7}	9.24
Median	6.57×10^{-8}	1.05×10^{-7}	8.12
1st Percentile	2.14×10^{-8}	5.05×10^{-8}	4.69
99th Percentile	2.28×10^{-7}	3.75×10^{-7}	
99.9th Percentile	3.84×10^{-7}	5.42×10^{-7}	
Reference range	$4.29–34.29 \times 10^{-8}$	$7.14–47.62 \times 10^{-8}$	
Affected range	<0.4	n/a	3.3

The effect of drying conditions on GAA enzyme activity was studied. Duplicate specimens were collected. One was dried at room temperature overnight (dry) while the other remained in a plastic wrapping for 2 days to simulate transport without proper drying (wrapped). After this time period the sample was taken out of the plastic bag and dried overnight. Both were tested using the fluorometric method with and without addition of acarbose and as at pH 7.0 (reference enzyme). A significant decrease of enzyme activity (on average 50%) was observed if samples were not dried properly. In two cases (sample 2 and 3) it led to results that could be interpreted as consistent with Pompe disease or carrier status, while the interpretation of sample 5 changed from borderline positive into an unsuitable specimen. This observation is consistent with previously reported results [30]. Data are summarized in Table 4.

Table 4. Degradation study of GAA activities for five dried blood specimens that showed different index activities. Samples were either dried overnight after spotting (dry) or put into sealed plastic bags immediately, in order to simulate shipping without proper drying (wrapped). After 2 days the samples were taken out and allowed to dry overnight. Both specimens were tested in the same run in duplicates. All samples showed a significant decrease in activity when the specimen was not dried before shipping.

α-Glucosidase Activity [nmol/punch × 21 h] (Reference Range)	Sample 1		Sample 2		Sample 3		Sample 4		Sample 5	
	Wrapped	Dry	Wrapped	Dry	Wrapped	Dry	Wrapped	Dry	Wrapped	Dry
pH 3.8 (>1.5)	2.07	3.02	0.82	2.06	1.28	3.09	1.92	2.56	0.29	1.6
pH 7.0 (>1.8)	5.4	6.47	1.8	4.48	3.33	6.12	4.07	6.04	0.9	5.52
pH 3.8 with acarbose (>0.9)	1	1.72	0.54	1.19	0.72	1.86	1.12	1.35	0.18	0.58

4. Discussion

Newborn screening for Pompe disease can be performed by a variety of methods based on enzyme activity measurement. NBS provides early identification of both classic severe IOPD and less-severe LOPD patients. With early detection and ERT, there are benefits for classic severe IOPD patients, however current therapy has limitations, especially with respect to the neurologic manifestations of the disease. The combination of early detection, close monitoring, and early ERT may be beneficial to less-severe LOPD patients as well. However, in some countries and regions the identification of LOPD presents an ethical problem and whole population screening poses financial constraints on health care systems. In contrast, at-risk testing as presented here, may be a potential first step. It allows for earlier identification of IOPD cases and potentially also LOPD when performed with a targeted approach using nonspecific symptoms loosely associated with Pompe disease, such as CK elevations of unknown origin. Interestingly, usually mild to moderate CK elevations are observed in patients with Pompe disease, which do not improve under therapy. In our study we have received specimens from patients who presented with non-specific symptoms and thus, may have Pompe disease but may have had another disorder hence GAA measurement aids in the differential diagnosis. Among 13,627 samples tested, 8% had decreased enzyme activity. We have shown that detrimental pre-analytical conditions may also affect enzyme activities negatively, so further confirmation is necessary. For that purpose, another enzyme measurement can be performed, and a molecular genetic assay should also be carried out. Previously, we have demonstrated that about. 2% of patients tested are eventually confirmed with Pompe disease which is in agreement with other international studies [31,32]. This demonstrates the high efficacy of the at-risk testing approach. The time to diagnosis can be significantly shortened, especially for LOPD patients who present with less specific symptoms. For IOPD patients, testing may be helpful in regions that are not familiar with the specific symptoms however due to the first clear symptoms, in particular floppiness, which occurs around 2–3 months of age, an even earlier diagnosis remains restricted to newborn screening.

Interest in Pompe disease testing within NBS programs has increased substantially in recent years. Sample quality greatly affects results from Pompe testing and newborn screening in general. As previously described [30], the combination of humidity and heat has the greatest impact on enzyme stability. The authors also showed with their shipping study that when properly dried DBS were stored in either paper envelopes or plastic bags, enzyme activities remained essentially intact for nine days in the US postal setting [30]. In contrast, insufficient drying combined with shipping of the samples in sealed (plastic) containers leads to grossly reduced activity, especially of the acid α-glucosidase and may even result in erroneous interpretation of the results as the activity of the reference enzyme remains in the normal range.

For adequate samples, GAA levels in specimens from affected patients are well resolved from those observed in specimens from healthy subjects using either fluorometry or tandem mass spectrometry (Table 3). The strength of mass spectrometry lies especially in its ability to measure several enzyme activities simultaneously. This is beneficial for newborn screening when various different lysosomal storage disorders are included in a national panel. Furthermore, the different enzyme activities may aid in the evaluation of sample quality and thus differentiate the cause of low enzyme activity between deficiency in the individual and deficiency caused by inappropriate storage conditions or transport. Using the fluorometric method as described the activity of α-glucosidase at pH 7.0 can be measured. This is usually less stable than the acid α-glucosidase and therefore, is a good indicator of negative environmental influences. For α-glucosidase tests which only include the activity with inhibition, β-galactosidase may be an alternative reference enzyme to monitor sample quality. This additional fluorometric test is fast and inexpensive, however, it must be used with some caution. β-galactosidase in dried blood may be relatively stable, therefore, a low enzyme activity result for α-glucosidase may still be caused by pre-analytical conditions rather than an actual disease. This applies to reference enzymes in general, as different environmental conditions may affect these enzymes to varying degrees. Thus, despite problematic conditions prior to the arrival of the specimen in the laboratory, the reference enzyme(s) may still show normal activity levels in the DBS. Therefore, the assessment of another specimen, either again in DBS or in a different material should always be considered as a second step. For further confirmation, a molecular genetic assessment is necessary. In case of IOPD, it may replace the second enzyme assessment or can be carried out in parallel to save valuable time and initiate therapy more rapidly.

In conclusion, at-risk testing can provide a good first step for the future introduction of Pompe disease to a newborn screening program as the laboratory can introduce the required methodology and gain insights in the evaluation of results in a lower throughput environment. It yields immediate benefits for the patients regarding availability and timeliness of the diagnosis. Finally, awareness of this rare condition is increased tremendously among local physicians which can aid in the introduction of Pompe disease into a national newborn screening program.

Author Contributions: Conceptualization, Z.L. and D.C.K.; Data curation, Z.L., P.O., P.N.C., and T.P.M.; Investigation, T.P.M.; Project administration, Z.L.; Resources, D.C.K.; Validation, P.N.C. and J.S.; Writing—original draft, Z.L. and P.O.; Writing—review & editing, P.N.C., J.S., and D.C.K. All authors have read and agreed to the published version of the manuscript.

Funding: Funding for this study was provided by Sanofi-Genzyme and from the internal resources of ARCHIMEDlife.

Acknowledgments: We thank our colleagues from ARCHIMEDlife and Hamburg University Medical Center, who provided insight and expertise that greatly assisted with all the interpretations and conclusions of this paper.

Conflicts of Interest: The authors declare that there is no conflict of interest. The funders had no role in the design of the study; in the collection, analyses, or interpretation of data; in the writing of the manuscript, or in the decision to publish the results.

References

1. Chien, Y.-H.; Hwu, W.-L.; Lee, N.-C. Pompe disease: Early diagnosis and early treatment make a difference. *Pediatr. Neonatol.* **2013**, *54*, 219–227. [CrossRef]
2. Dasouki, M.; Jawdat, O.; Almadhoun, O.; Pasnoor, M.; McVey, A.L.; Abuzinadah, A.; Herbelin, L.; Barohn, R.J.; Dimachkie, M.M. Pompe disease: Literature review and case series. *Neurol. Clin.* **2014**, *32*, 751–776. [CrossRef]
3. Kohler, L.; Puertollano, R.; Raben, N. Pompe Disease: From Basic Science to Therapy. *Neurotherapeutics* **2018**, *15*, 928–942. [CrossRef]
4. Bay, L.B.; Denzler, I.; Durand, C.; Eiroa, H.; Frabasil, J.; Fainboim, A.; Maxit, C.; Schenone, A.; Spécola, N. Infantile-onset Pompe disease: Diagnosis and management TT—Enfermedad de Pompe infantil: Diagnóstico y tratamiento. *Arch. Argent. Pediatr.* **2019**, *117*, 271–278.
5. Van den Hout, H.M.P.; Hop, W.; van Diggelen, O.P.; Smeitink, J.A.M.; Smit, G.P.A.; Poll-The, B.-T.T.; Bakker, H.D.; Loonen, M.C.B.; de Klerk, J.B.C.; Reuser, A.J.J.; et al. The natural course of infantile Pompe's disease: 20 original cases compared with 133 cases from the literature. *Pediatrics* **2003**, *112*, 332–340. [CrossRef] [PubMed]

6. Case, L.E.; Beckemeyer, A.A.; Kishnani, P.S. Infantile Pompe disease on ERT: Update on clinical presentation, musculoskeletal management, and exercise considerations. *Am. J. Med. Genet. C Semin. Med. Genet.* **2012**, *160*, 69–79. [CrossRef]

7. Cupler, E.J.; Berger, K.I.; Leshner, R.T.; Wolfe, G.I.; Han, J.J.; Barohn, R.J.; Kissel, J.T.; AANEM Consensus Committee on Late-onset Pompe Disease. Consensus treatment recommendations for late-onset Pompe disease. *Muscle Nerve* **2012**, *45*, 319–333. [CrossRef] [PubMed]

8. Kishnani, P.S.; Hwu, W.-L.; Group, P.D.N.S.W. Introduction to the Newborn Screening, Diagnosis, and Treatment for Pompe Disease Guidance Supplement. *Pediatrics* **2017**, *140*, S1–S3. [CrossRef]

9. Metz, T.F.; Mechtler, T.P.; Orsini, J.J.; Martin, M.; Shushan, B.; Herman, J.L.; Ratschmann, R.; Item, C.B.; Streubel, B.; Herkner, K.R.; et al. Simplified newborn screening protocol for lysosomal storage disorders. *Clin. Chem.* **2011**, *57*, 1286–1294. [CrossRef]

10. Mechtler, T.P.; Stary, S.; Metz, T.F.; De Jesus, V.R.; Greber-Platzer, S.; Pollak, A.; Herkner, K.R.; Streubel, B.; Kasper, D.C. Neonatal screening for lysosomal storage disorders: Feasibility and incidence from a nationwide study in Austria. *Lancet* **2012**, *379*, 335–341. [CrossRef]

11. Bali, D.S.; Goldstein, J.L.; Banugaria, S.; Dai, J.; Mackey, J.; Rehder, C.; Kishnani, P.S. Predicting cross-reactive immunological material (CRIM) status in Pompe disease using GAA mutations: Lessons learned from 10 years of clinical laboratory testing experience. *Am. J. Med. Genet. C Semin. Med. Genet.* **2012**, *160C*, 40–49. [CrossRef]

12. Kishnani, P.S.; Corzo, D.; Leslie, N.D.; Gruskin, D.; Van der Ploeg, A.; Clancy, J.P.; Parini, R.; Morin, G.; Beck, M.; Bauer, M.S.; et al. Early treatment with alglucosidase alpha prolongs long-term survival of infants with Pompe disease. *Pediatr. Res.* **2009**, *66*, 329–335. [CrossRef]

13. Wang, Z.; Okamoto, P.; Keutzer, J. A new assay for fast, reliable CRIM status determination in infantile-onset Pompe disease. *Mol. Genet. Metab.* **2014**, *111*, 92–100. [CrossRef]

14. Chien, Y.-H.; Hwu, W.-L.; Lee, N.-C. Newborn screening: Taiwanese experience. *Ann. Transl. Med.* **2019**, *7*, 281. [CrossRef]

15. Elliott, S.; Buroker, N.; Cournoyer, J.J.; Potier, A.M.; Trometer, J.D.; Elbin, C.; Schermer, M.J.; Kantola, J.; Boyce, A.; Turecek, F.; et al. Pilot study of newborn screening for six lysosomal storage diseases using Tandem Mass Spectrometry. *Mol. Genet. Metab.* **2016**, *118*, 304–309. [CrossRef]

16. Bodamer, O.A.; Scott, C.R.; Giugliani, R.; Pompe Disease Newborn Screening Working Group. Newborn Screening for Pompe Disease. *Pediatrics* **2017**, *140*, S4–S13. [CrossRef]

17. Wasserstein, M.P.; Caggana, M.; Bailey, S.M.; Desnick, R.J.; Edelmann, L.; Estrella, L.; Holzman, I.; Kelly, N.R.; Kornreich, R.; Kupchik, S.G.; et al. The New York pilot newborn screening program for lysosomal storage diseases: Report of the first 65,000 infants. *Genet. Med.* **2019**, *21*, 631–640. [CrossRef]

18. Chamoles, N.A.; Niizawa, G.; Blanco, M.; Gaggioli, D.; Casentini, C. Glycogen storage disease type II: Enzymatic screening in dried blood spots on filter paper. *Clin. Chim. Acta* **2004**, *347*, 97–102. [CrossRef]

19. Zhang, H.; Kallwass, H.; Young, S.P.; Carr, C.; Dai, J.; Kishnani, P.S.; Millington, D.S.; Keutzer, J.; Chen, Y.-T.; Bali, D. Comparison of maltose and acarbose as inhibitors of maltase-glucoamylase activity in assaying acid alpha-glucosidase activity in dried blood spots for the diagnosis of infantile Pompe disease. *Genet. Med.* **2006**, *8*, 302–306. [CrossRef]

20. Niizawa, G.; Levin, C.; Aranda, C.; Blanco, M.; Chamoles, N.A. Retrospective diagnosis of glycogen storage disease type II by use of a newborn-screening card. *Clin. Chim. Acta* **2005**, *359*, 205–206. [CrossRef]

21. Sista, R.S.; Wang, T.; Wu, N.; Graham, C.; Eckhardt, A.; Winger, T.; Srinivasan, V.; Bali, D.; Millington, D.S.; Pamula, V.K. Multiplex newborn screening for Pompe, Fabry, Hunter, Gaucher, and Hurler diseases using a digital microfluidic platform. *Clin. Chim. Acta* **2013**, *424*, 12–18. [CrossRef]

22. Spáčil, Z.; Elliott, S.; Reeber, S.L.; Gelb, M.H.; Scott, C.R.; Tureček, F. Comparative triplex tandem mass spectrometry assays of lysosomal enzyme activities in dried blood spots using fast liquid chromatography: Application to newborn screening of Pompe, Fabry, and Hurler diseases. *Anal. Chem.* **2011**, *83*, 4822–4828. [CrossRef]

23. Mechtler, T.P.; Metz, T.F.; Muller, H.G.; Ostermann, K.; Ratschmann, R.; De Jesus, V.R.; Shushan, B.; Di Bussolo, J.M.; Herman, J.L.; Herkner, K.R.; et al. Short-incubation mass spectrometry assay for lysosomal storage disorders in newborn and high-risk population screening. *J. Chromatogr. B Anal. Technol. Biomed. Life Sci.* **2012**, *908*, 9–17. [CrossRef] [PubMed]

24. Xia, B.; Asif, G.; Arthur, L.; Pervaiz, M.A.; Li, X.; Liu, R.; Cummings, R.D.; He, M. Oligosaccharide analysis in urine by maldi-tof mass spectrometry for the diagnosis of lysosomal storage diseases. *Clin. Chem.* **2013**, *59*, 1357–1368. [CrossRef]

25. Young, S.P.; Piraud, M.; Goldstein, J.L.; Zhang, H.; Rehder, C.; Laforet, P.; Kishnani, P.S.; Millington, D.S.; Bashir, M.R.; Bali, D.S. Assessing disease severity in Pompe disease: The roles of a urinary glucose tetrasaccharide biomarker and imaging techniques. *Am. J. Med. Genet. C Semin. Med. Genet.* **2012**, *160C*, 50–58. [CrossRef]

26. Chien, Y.-H.; Goldstein, J.L.; Hwu, W.-L.; Smith, P.B.; Lee, N.-C.; Chiang, S.-C.; Tolun, A.A.; Zhang, H.; Vaisnins, A.E.; Millington, D.S.; et al. Baseline Urinary Glucose Tetrasaccharide Concentrations in Patients with Infantile- and Late-Onset Pompe Disease Identified by Newborn Screening. *JIMD Rep.* **2015**, *19*, 67–73. [PubMed]

27. Lukacs, Z.; Nieves Cobos, P.; Mengel, E.; Hartung, R.; Beck, M.; Deschauer, M.; Keil, A.; Santer, R. Diagnostic efficacy of the fluorometric determination of enzyme activity for Pompe disease from dried blood specimens compared with lymphocytes-possibility for newborn screening. *J. Inherit. Metab. Dis.* **2010**, *33*, 43–50. [CrossRef]

28. Verma, J.; Thomas, D.C.; Kasper, D.C.; Sharma, S.; Puri, R.D.; Bijarnia-Mahay, S.; Mistry, P.K.; Verma, I.C. Inherited Metabolic Disorders: Efficacy of Enzyme Assays on Dried Blood Spots for the Diagnosis of Lysosomal Storage Disorders. *JIMD Rep.* **2017**, *31*, 15–27.

29. Liao, H.-C.; Chiang, C.-C.; Niu, D.-M.; Wang, C.-H.; Kao, S.-M.; Tsai, F.-J.; Huang, Y.-H.; Liu, H.-C.; Huang, C.-K.; Gao, H.-J.; et al. Detecting multiple lysosomal storage diseases by tandem mass spectrometry—A national newborn screening program in Taiwan. *Clin. Chim. Acta* **2014**, *431*, 80–86. [CrossRef]

30. Elbin, C.S.; Olivova, P.; Marashio, C.A.; Cooper, S.K.; Cullen, E.; Keutzer, J.M.; Zhang, X.K. The effect of preparation, storage and shipping of dried blood spots on the activity of five lysosomal enzymes. *Clin. Chim. Acta* **2011**, *412*, 1207–1212. [CrossRef] [PubMed]

31. Lukacs, Z.; Cobos, P.N.; Wenninger, S.; Willis, T.A.; Guglieri, M.; Roberts, M.; Quinlivan, R.; Hilton-Jones, D.; Evangelista, T.; Zierz, S.; et al. Prevalence of Pompe disease in 3,076 patients with hyperCKemia and limb-girdle muscular weakness. *Neurology* **2016**, *87*, 295–298. [CrossRef] [PubMed]

32. Lukacs, Z.; Schoser, B. Meta-opinion: From screening to diagnosis of Pompe disease—A European perspective. *Expert Opin. Orphan Drugs* **2016**, *4*, 1075–1078. [CrossRef]

Publisher's Note: MDPI stays neutral with regard to jurisdictional claims in published maps and institutional affiliations.

International Journal of
Neonatal Screening

Case Report

A Newborn Screening, Presymptomatically Identified Infant With Late-Onset Pompe Disease: Case Report, Parental Experience, and Recommendations

Raymond Y. Wang [1,2,†]

1   Division of Metabolic Disorders, CHOC Children's Specialists, Orange, CA 92868, USA; rawang@choc.org
2   Department of Pediatrics, University of California-Irvine School of Medicine, Orange, CA 92868, USA
†   The parents of the proposita presented firmly state their desire to maintain anonymity. However, they have
    provided written consent for publication of the proposita's details and their commentary.

Received: 22 January 2020; Accepted: 12 March 2020; Published: 14 March 2020

Abstract: Pompe disease is an inherited lysosomal storage disorder caused by acid alpha-glucosidase (GAA) enzyme deficiency, resulting in muscle and neuron intralysosomal glycogen storage. Clinical symptoms vary from the severe, infantile-onset form with hypertrophic cardiomyopathy, gross motor delay, and early death from respiratory insufficiency; to a late-onset form with variable onset of proximal muscle weakness and progressive respiratory insufficiency. Newborn screening programs have been instituted to presymptomatically identify neonates with infantile-onset Pompe disease for early initiation of treatment. However, infants with late-onset Pompe disease are also identified, leaving families and physicians in a state of uncertainty regarding prognosis, necessity, and timing of treatment initiation. This report presents a 31 5/7 weeks' gestational age premature infant flagged positive for Pompe disease with low dried blood spot GAA activity; sequencing identified biparental c.-32-13T>G/c.29delA *GAA* variants predicting late-onset Pompe disease. The infant's parents' initial reactions to the positive newborn screen, subsequent experience during confirmatory testing, and post-confirmation reflections are also reported. While uncertainties regarding natural history and prognosis of presymptomatically-identified late-onset Pompe disease infants will be elucidated with additional experience, suggestions for education of first-line providers are provided to accurately communicate results and compassionately counsel families regarding anxiety-provoking positive newborn screen results.

Keywords: Pompe disease; late-onset; infantile-onset; newborn screening; presymptomatic; c.-32-13T>G

---

1. Introduction

Pompe disease, caused by acid α-glucosidase enzyme deficiency (GAA; EC 3.2.1.20) due to pathogenic variants in *GAA*, is characterized by intralysosomal accumulation of glycogen throughout bodily tissues, most notably within cardiac and skeletal muscle.

Muscle lysosomal glycogen storage results in hypertrophic cardiomyopathy and skeletal muscle weakness varying in age of onset and severity according to residual GAA enzymatic activity. Skeletal myopathy is also augmented by spinal cord anterior horn and brain neuronal glycogen storage and dysfunction [1]. Infantile-onset Pompe disease (IOPD), caused by near-absence of lysosomal GAA enzyme, typically manifests in the first two months of life with progressive and severe hypertrophic cardiomyopathy, heart failure, marked hypotonia, respiratory failure; if untreated, death from cardiopulmonary complications typically occurs within the first 14 months of life [2].

Late-onset Pompe disease (LOPD) patients have pathogenic *GAA* variants that reduce, but do not completely abolish, acid α-glucosidase enzyme activity. The residual enzyme attenuates the disease course; prior to initiation of newborn screening programs for Pompe disease, LOPD patients were often diagnosed in adulthood following decades of proximal myopathic symptoms. Without treatment, LOPD patients experience deterioration of skeletal muscle strength, becoming nonambulatory and dependent upon artificial ventilation (non-invasive positive airway pressure, or invasive mechanical ventilation) [3].

Pompe disease patients typically have elevations of serum creatine phosphokinase and transaminases secondary to myopathic injury. However, these tests are neither sensitive nor specific for Pompe disease, nor is urinary hexose tetrasaccharide, which is an excreted marker of glycogen accumulation. A diagnosis of Pompe disease relies upon identification of deficient white blood cell GAA enzymatic activity and biparentally inherited variants in *GAA* [4].

Development of intravenous enzyme replacement therapy (ERT) in the form of recombinant human GAA (rhGAA) enzyme, has radically transformed the natural history of Pompe disease. The efficacy of rhGAA ERT is well documented for both IOPD and LOPD. IOPD patients' overall survival and ventilation-free survival are drastically prolonged; early initiation of rhGAA therapy is associated with better developmental and survival outcomes [5]. Since immune response and cross-reactive immunologic material (CRIM) status play a significant role in outcomes, pre-rhGAA immunomodulation is also indicated to mitigate the neutralizing effect of the anti-rhGAA immune response [6]. LOPD patients also demonstrate improvement in ambulatory and pulmonary function, with delayed onset of declines in six-minute walk and pulmonary function testing [7].

Given the significant improvement in survival, hypertrophic cardiomyopathy, and pulmonary outcomes in infantile-onset patients, Pompe disease was included in newborn screening (NBS) programs around the world. Taiwan, which has been screening for Pompe disease since 2005 due to a founder effect for IOPD [8], reported 28 newborns affected with either form of Pompe disease out of 473,738 screened (1 case per 16,919 newborns) [9]. Surprisingly, prevalences from NBS programs without known Pompe disease founder effects have also been much higher than expected. The Austrian program identified 4 Pompe disease infants out of 34,736 screened (1 case per 8,684 newborns) [10]. Pilot NBS programs in the United States also identified a high prevalence of Pompe disease; the state of Missouri identified 8 affected out of 43,701 screened (1 case per 5463 newborns) [11], while the state of Illinois identified 10 affected out of 219,973 screened (1 case per 21,979 newborns) [12].

Favorable outcomes of NBS-identified IOPD infants in these programs led to the eventual inclusion of Pompe disease to the United States Health and Human Services "Recommended Uniform Screening Panel" in 2013, leading to the implementation of universal newborn screening for Pompe disease in the United States. NBS for Pompe disease began in the state of California in July of 2018.

New challenges emerge with the implementation of every newborn screening program. Quantification of GAA enzymatic activity in NBS samples cannot distinguish between IOPD and LOPD patients. Consequently, for each IOPD case identified, there are anywhere from 1 to 4 LOPD cases identified. If GAA enzymatic activity alone is utilized as the first-tier analyte, heterozygous and pseudodeficiency-carrying infants are also identified. Though rapid initiation of rhGAA ERT is clearly beneficial for IOPD babies, the management—including potential decisions regarding ERT—of NBS-identified, presymptomatic LOPD infants is not clearly defined [13].

Because NBS for Pompe disease was recently introduced in the United States, a clearer understanding of early childhood LOPD phenotypes and natural history will require more time for additional cases to be identified and followed longitudinally. Only then can clear guidelines for clinical follow-up and potential initiation of ERT be issued.

These early uncertainties present difficulties for the clinicians who are the initial evaluators of NBS-identified LOPD infants, and create anxiety and distress for the infants' parents. This scenario is not novel to the practice of newborn screening: Following implementation of universal screening for phenylketonuria, another inborn error of metabolism for which early institution of treatment significantly impacts outcome, parents of infants ultimately found to be false positives for the disorder demonstrated long-lasting anxiety about their baby's health [14]. In the era of tandem mass spectrometry-based NBS for multiple inborn errors of metabolism, infants with NBS false positives were hospitalized twice as often as infants with normal NBS [15] due to higher parental stress levels, dysfunction, and perceptions of fragility. If such a possibility for psychological harm exists for parents of infants with false positives, it is undoubtedly present for parents of infants with attenuated conditions for which no clear standard of care yet exists.

The purpose of this report is to highlight one case of LOPD identified through the California Newborn Screening program. More importantly, this report aims to gain insight into how the infant's parents initially responded to the revelation of their child's condition, and how they processed and coped with the realization that their child had LOPD. Their goal in sharing their experience and suggestions, which have been curated, is to help all medical staff involved in the newborn screening process—not only medical geneticists, but also the primary care pediatricians' offices and neonatal intensive care units—better understand the complexities of screening for a heterogeneous disorder and how to effectively counsel and communicate with families whose infants have been identified by NBS.

2. Case Report

Parents of the proposita have provided consent for publication of clinical data. The proposita is a second twin, born at 31 5/7 weeks' gestational age with a birth weight of 1550 g. Her mother was a 36-year-old gravida 2/para 1-2-0-3 woman whose twin pregnancy was monitored closely due to anti-protein S antibodies; she experienced prolonged premature rupture of membranes leading to the proposita's delivery via repeat Cesarean section. Prenatal screening included normal prenatal nuchal translucency, first trimester laboratory testing, and monitoring ultrasounds.

Initial Apgar scores were 7 at 1 min and 8 at 5 min. She experienced an uncomplicated neonatal course; in the perinatal period, she received empiric antibiotics, phototherapy, and was placed on nasal continuous positive airway pressure. She did not require intubation or supplemental oxygen, and was hospitalized for a total of four weeks in the neonatal intensive care unit advancing oral feedings.

Her newborn screening was flagged for potential Pompe disease as blood spot GAA enzymatic activity was low at 2.7 nmol/mg protein/h (reference range 20.9–140.7). Control β-galactosidase enzyme was within normal range. Blood creatine phosphokinase level was 134 units/liter (reference range <235 units/liter). While urinary hexose tetrasaccharide (uHex4) and *GAA* molecular sequencing were pending, an echocardiogram was performed which documented normal atrial and ventricular dimensions, normal left ventricular wall thickness without evidence of cardiomyopathy, and a normal left ventricular ejection fraction of 72.4%.

Initial uHex4 level was elevated at 25.2 mmol/mol creatinine (reference range <17.6 mmol/mol creatinine); because this was obtained when she was 32 5/7 weeks' adjusted gestational age, when low urinary creatinine excretion of a premature infant could interfere with accurate measurement, it was repeated five weeks later which was normal at 4.1 mmol/mol creatinine. *GAA* sequencing identified two pathogenic variants: the paternally inherited "common" late-infantile onset c.-32-13T>G variant, and the maternally inherited c.29delA (p.His10Profs*33) variant.

Subsequent measurements of creatine phosphokinase continued to be normal at 215 units/L (23 days of life) and 212 units/L (2.5 months of life). An echocardiogram performed at two months of life continued to show normal ventricular dimensions and function without left ventricular hypertrophy. She had retinopathy of prematurity, which subsequently resolved. She also had apnea of prematurity, for which she was initially placed on caffeine citrate while hospitalized, and subsequently monitored with an apnea monitor after two episodes of apnea following discharge.

Prior to her first evaluation with the author, she was rehospitalized twice for respiratory issues. The first, occurring at 6 weeks of age, was due to respiratory distress caused by rhinovirus bronchiolitis, for which she was placed on nasal continuous positive airway pressure with supplemental oxygen. The second, occurring at 8 weeks of age, took place because she experienced apnea at home requiring cardiopulmonary resuscitation. She was hospitalized for intravenous antibiotics for treatment of presumptive pneumonia and for observation.

Following her hospitalizations, she experienced good subsequent weight gain without any feeding difficulties. At approximately four months of age (two months adjusted for prematurity), her weight was 5.13 kg (50th percentile), length was 54.6 cm (15th percentile), and head circumference 39.5 cm (75th percentile). Examination identified head lag, absence of calf pseudohypertrophy, normal patellar/biceps/triceps/brachioradialis deep tendon reflexes, and reduced truncal and appendicular muscle tone.

3. Parental Perspective

We had finally escaped the fluorescent lighting and beeps of the machines in the NICU for a brief moment to take our older son and my mother out for a well-needed frozen yogurt. The phone rang—with the number of the hospital on the screen—sending chills down my spine. The doctor informed us that one of our twins, who were born two months prematurely, had an abnormality on the newborn screening test. "What disease and what baby?" I asked.

"Pompeii disease in baby B" he seemed to have said. Images of an entire civilization frozen in molten lava came flooding through my mind. "It's a metabolic disease. It's most likely a false positive. We see this a lot. Once the baby eats and the metabolism stabilizes, the test will most likely be normal. We will confirm with a genetic test and a few other tests. Just don't look it up online." So, of course, immediately, my husband and I looked it up. Autocorrect changed "Pompeii" to "Pompe." From the (now I understand outdated) catastrophic and deadly description of the disease online, I would have preferred a Roman volcanic eruption.

We would later learn that this description was for IOPD before the current enzyme replacement therapy treatments, which have dramatically improved the outcome for Pompe disease patients. We would also later learn that there are at least two very distinct types of Pompe disease, infantile, and late-onset, and that doctors, researchers, and biotech companies have made tremendous strides in understanding, treating, and halting the progression of the disease. Some born with IOPD can walk, dance, sing, and even attend college. Late-onset patients are now identified more quickly due to newborn screening, and can be treated before muscular deterioration occurs. I have since met a successful scientist, a neurologist, and the head of a foundation who all have LOPD, and you would most likely never know they had it. Furthermore, with gene therapy clinical trials already accepting applicants and gene editing on the horizon, there is even the prospect of the eliminating the disease entirely.

However, at this point, we were only seeing what was online. "It's a lysosomal storage disease", I said to my mother while driving back in the car. She stopped the car, paralyzed. "That's what your brother had. It's a different gene but the same category of disease." How could this be? My brother had died of Batten Disease, and my husband and I underwent extensive genetic counseling to help insure that none of our children would have a rare genetic disease.

What I found out later about prenatal carrier testing was that I was only tested for the common variant of Pompe disease (c.-32-13T>G), which I do not carry, but my husband does. I later discovered that I carry a previously unrecorded, but most likely very severe, mutation that contains a deletion. In our prenatal counseling, my husband was not tested as a carrier for Pompe disease at all. One counselor had told me I had the same risk as the general population. Another doctor said it would be unlikely that I would carry Pompe disease because I was not of Dutch descent.

When we did receive the genetic confirmation that our baby did indeed have the disease, the genotype–phenotype correlation baffled the NICU doctors and the consulting geneticist. She had the common variant (predicting late-onset Pompe disease) and this unrecorded severe mutation (predicting infantile Pompe disease). Whereas classic IOPD, defined by cardiomyopathy as well as low GAA enzymatic levels in an infant younger than one year of age, is very severe, LOPD can be potentially mild with a later multi-symptom disease progression. Cardiac involvement is rarely involved in the c.-32-13T>G mutation [16]. In one study, ERT was initiated early with some LOPD infants exhibiting very mild early signs of muscle weakness [17].

However, because one of our child's confirmatory tests were abnormal with elevated uHex4 levels, which could indicate glycogen storage, the doctors did not know whether the baby had infantile or late-onset Pompe disease. Although infantile Pompe disease is sometimes known as "floppy baby syndrome," she was not floppy at all but had excellent muscle tone. Her treatment depended upon a correct diagnosis: infantile-onset Pompe disease needed enzyme replacement therapy treatment right away, while for late-onset Pompe disease, enzyme replacement therapy was less urgent. The laboratory performing the uHex4 test informed us that this elevation was due to prematurity; thankfully, in the next test, the uHex4 levels were normal.

There was quite a learning curve with terms like "uHex4," "ERT (enzyme replacement therapy)", "leaky splice site," "c.-32-13T>G," "GAA-levels", "CK levels," "exon," "del (deletion)," and "dup (duplication)", but after a few months of research, we found ourselves able to understand much of the genetic lingo and code. I found it most helpful to start with the articles referencing the specific alleles (if previously reported), which were cited in the genetic confirmation of the disease. This evolved to understanding broader issues and concepts in the Pompe community by attending conferences focusing on Pompe or rare disease and by making contact with leading researchers, doctors, and biotech companies interested in Pompe as well as parents of kids with Pompe and patients with Pompe.

We also researched observational studies and clinical trials. There are many groups on social media to connect people as well. In the rare disease community, we discovered, as one patient put it, "patients and parents are treated as collaborators." There is so much hope in this field, and no parent needs to be devastated by the newborn screening results. In fact, we are also filled with gratitude that this disease was caught early and can be treated and eventually eliminated.

4. Suggestions from Parents, to Medical Staff Counseling or Seeing Families Screening Positive for Pompe NBS

4.1. Screening and Genetic Counseling

Do not minimize the initial results for the parents by telling them that chances are the test is a false positive. The follow up testing may in fact demonstrate that the initial result was in fact a true positive.

Do not tell parents that infantile Pompe disease means that the child will die as an infant.

Do not tell parents not to look up the disease online. Instead, give them suggestions of what to look up and what to ignore.

A person does not have to be Dutch to carry a Pompe disease gene mutation.

A baby that "looks healthy" can still have a genetic disease.

The common c.-32-13T>G variant of Pompe, even when present on one allele with a normal echo, does not result in IOPD. It predicts later onset Pompe disease.

4.2. Pompe Disease Management and Treatment

Do not early discharge premature infants from the NICU if Pompe disease, or any other serious metabolic disorder, is suspected. The geneticist should perform a clinical evaluation and examination of the child. Prematurity should also be closely monitored.

Teach other health care staff about the difference between infantile and late-onset Pompe disease.

Learn about enzyme replacement therapy and inform the parents about it.

IOPD babies must be treated with enzyme replacement therapy as soon as possible once CRIM status has been confirmed and immunomodulation potentially initiated [6].

Give a referral to a dietician versed in Pompe disease. Diets should be high in protein but low in simple carbohydrates [18].

Give a referral to physical therapy if possible and if necessary. We have found that physical therapy can help build up muscle strength even without ERT yet.

4.3. Family Advocacy

Tell the parents about all the incredible resources there are for Pompe patients, which will inform parents of gene therapy studies, observational newborn studies, clinical trials, next generation ERT therapies, pharmacological chaperones and put them in touch with the Pompe community.

Tell parents to research and contact the leading Pompe doctors, researchers, and biotech companies in the country.

Inform the parents about registries.

Tell the parents to have hope! Pompe disease has been thoroughly researched. It is precisely because there is treatment that it is on the NBS, and the one word that continuously circulates in discussions of upcoming and eventual gene therapies for Pompe disease is "cure."

5. Discussion

This report highlights a case of a neonate in the State of California screening positive for Pompe disease. Reported herein are findings of clinical, biochemical, and molecular confirmatory testing that ultimately identified a diagnosis of late-onset Pompe disease. Additionally reported are the reactions and emotional journey taken by the infant's parents from initial notification, to definitive diagnosis, and subsequent follow-up.

The child's neonatal course was complicated by prematurity, not only due to premature physiology resulting in increased excretion of urinary hexose tetrasaccharide, but also due to apnea of prematurity episodes/intercurrent illness in which she required supplemental oxygen or cardiopulmonary resuscitation. She has hypotonia on physical examination, and it is too early to tell if the hypotonia is a result of her prematurity, potential neurologic sequelae from hypoxia, or due to early symptoms of LOPD.

Review of clinically-diagnosed (not NBS-identified) Pompe disease patients indicates that patients with the common c.-32-13T>G intronic splice site *GAA* variant uniformly develop LOPD are very rarely develop childhood-onset cardiac disease [16]. While most of these patients develop myopathic symptoms and are at risk of dysrhythmias and left ventricular hypertrophy in adulthood, a small subset may manifest gross motor delay and proximal muscle weakness during their first years of life. A retrospective review of 84 symptomatically-identified LOPD patients with at least one c.-32-13T>G variant identified four patients presenting with symptoms prior to 20 months of age; of these patients, three had elevated creatine phosphokinase and two had elevated uHex4 levels [19]. An examination of 7 NBS-identified LOPD patients with the c.-32-13T>G variant demonstrated that while all infants had some degree of muscle weakness, the four patients with the most notable weakness had elevations in creatine phosphokinase. All NBS-identified infants had normal uHex4 levels [17].

For the infant in this report, physical therapy was recommended due to her hypotonia, prematurity, apnea of prematurity, and LOPD diagnosis. Quarterly monitoring for signs and symptoms of spinal and pelvic girdle muscle weakness, calf pseudohypertrophy, as well as laboratory quantification of transaminases, creatine phosphokinase, and uHex4 was instituted. Given the difficulties of secure intravenous access, potential for infusion-associated reactions, and absence of creatine phosphokinase/uHex4 elevation at assessment, ERT was deferred. The infant is now considered a "patient-in-waiting", with her parents and her metabolic physician placing her under surveillance for development of symptoms, "living between health and disease" and unable to determine when treatment should be instituted [20].

Pompe disease was added to newborn screening panels in order to rapidly identify and initiate treatment for neonates with the severe, infantile-onset form of the disease. Cardiac, motor, and overall survival outcomes in early-treated IOPD babies is superior to outcomes in IOPD babies initiated with ERT after symptomatic presentation [13]. However, since NBS identifies neonates with LOPD as well as those bearing pseudodeficiency variants and *GAA* variant heterozygotes, the implementation of NBS for Pompe disease has created considerable uncertainty. First, for parents of NBS-flagged infants, who must wait for results of confirmatory testing and undoubtedly experience some degree of anxiety after reading about the morbidity and mortality associated with Pompe disease; second, for medical providers—especially primary care practitioners and non-physician staff—who are often tasked with notifying parents about NBS results, but may feel their knowledge about Pompe disease is inadequate and subsequently either minimize the significance of the NBS or transmit their apprehension to the parents. Finally, after confirmation of LOPD, uncertainty exists for parents and specialty caregivers as there is currently no established management guideline for presymptomatically-identified LOPD infants. Our family's experience mirrors closely those of other families receiving Pompe disease NBS results [21]. As additional LOPD cases are identified and followed longitudinally with existing NBS-identified LOPD cases, a clearer natural history, prognosis, and recommendations for timing of treatment initiation will arise. Increased knowledge will help allay some of the anxieties and uncertainties faced by families of NBS-identified LOPD babies.

It is our parents' hope that their suggestions, which we are in agreement with, and the experience of other families will enhance health care practitioners' abilities to accurately communicate results and compassionately counsel families regarding anxiety-provoking positive newborn screen results.

Funding: R.Y.W. is supported by the Campbell Foundation of Caring and the Liferay Foundation.

Acknowledgments: The author is grateful to the patient and her family, for sharing their experiences.

Conflicts of Interest: The authors declare no conflict of interest.

References

1. Fuller, D.D.; ElMallah, M.K.; Smith, B.K.; Corti, M.; Lawson, L.A.; Falk, D.J.; Byrne, B.J. The respiratory neuromuscular system in Pompe disease. *Respir. Physiol. Neurobiol.* **2013**, *189*, 241–249. [CrossRef] [PubMed]
2. Kishnani, P.S.; Goldenberg, P.C.; DeArmey, S.L.; Heller, J.; Benjamin, D.; Young, S.; Bali, D.; Smith, S.A.; Li, J.S.; Mandel, H.; et al. Cross-reactive immunologic material status affects treatment outcomes in Pompe disease infants. *Mol. Genet. Metab.* **2010**, *99*, 26–33. [CrossRef] [PubMed]
3. Hagemans, M.L.; Winkel, L.P.; Van Doorn, P.A.; Hop, W.J.; Loonen, M.C.; Reuser, A.J.; Van der Ploeg, A.T. Clinical manifestation and natural course of late-onset Pompe's disease in 54 Dutch patients. *Brain* **2005**, *128*, 671–677. [CrossRef]
4. Bali, D.S.; Goldstein, J.L.; Banugaria, S.; Dai, J.; Mackey, J.; Rehder, C.; Kishnani, P.S. Predicting cross-reactive immunological material (CRIM) status in Pompe disease using, G.A.A mutations: Lessons learned from 10 years of clinical laboratory testing experience. *Am. J. Med. Genet. C Semin. Med. Genet.* **2012**, *160*, 40–49. [CrossRef] [PubMed]
5. Chien, Y.H.; Lee, N.C.; Chen, C.A.; Tsai, F.J.; Tsai, W.H.; Shieh, J.Y.; Huang, H.J.; Hsu, W.C.; Tsai, T.H.; Hwu, W.L. Long-term prognosis of patients with infantile-onset Pompe disease diagnosed by newborn screening and treated since birth. *J. Pediatr.* **2015**, *166*, 985–991. [CrossRef] [PubMed]
6. Desai, A.K.; Li, C.; Rosenberg, A.S.; Kishnani, P.S. Immunological challenges and approaches to immunomodulation in Pompe disease: A literature review. *Ann. Transl. Med.* **2019**, *7*, 285. [CrossRef]
7. Schoser, B.; Stewart, A.; Kanters, S.; Hamed, A.; Jansen, J.; Chan, K.; Karamouzian, M.; Toscano, A. Survival and long-term outcomes in late-onset Pompe disease following alglucosidase alfa treatment: A systematic review and meta-analysis. *J. Neurol.* **2017**, *264*, 621–630. [CrossRef]
8. Chien, Y.H.; Lee, N.C.; Huang, H.J.; Thurberg, B.L.; Tsai, F.J.; Hwu, W.L. Later-onset Pompe disease: Early detection and early treatment initiation enabled by newborn screening. *J. Pediatr.* **2011**, *158*, 1023–1027. [CrossRef]
9. Chiang, S.C.; Hwu, W.L.; Lee, N.C.; Hsu, L.W.; Chien, Y.H. Algorithm for Pompe disease newborn screening: Results from the Taiwan screening program. *Mol. Genet. Metab.* **2012**, *106*, 281–286. [CrossRef]
10. Mechtler, T.P.; Metz, T.F.; Müller, H.G.; Ostermann, K.; Ratschmann, R.; de Jesus, V.R.; Shushan, B.; Di Bussolo, J.M.; Herman, J.L.; Herkner, K.R.; et al. Short-incubation mass spectrometry assay for lysosomal storage disorders in newborn and high-risk population screening. *J. Chromatogr. B Anal. Technol. Biomed. Life Sci.* **2012**, *908*, 9–17. [CrossRef]

11. Hopkins, P.V.; Campbell, C.; Klug, T.; Rogers, S.; Raburn-Miller, J.; Kiesling, J. Lysosomal storage disorder screening implementation: Findings from the first six months of full population pilot testing in Missouri. *J. Pediatr.* **2015**, *166*, 172–177. [CrossRef] [PubMed]

12. Burton, B.K.; Charrow, J.; Hoganson, G.E.; Waggoner, D.; Tinkle, B.; Braddock, S.R.; Schneider, M.; Grange, D.K.; Nash, C.; Shryock, H.; et al. Newborn Screening for Lysosomal Storage Disorders in Illinois: The Initial 15-Month Experience. *J. Pediatr.* **2017**, *190*, 130–135. [CrossRef] [PubMed]

13. Kronn, D.F.; Day-Salvatore, D.; Hwu, W.L.; Jones, S.A.; Nakamura, K.; Okuyama, T.; Swoboda, K.J.; Kishnani, P.S. Pompe Disease Newborn Screening Working Group. Management of Confirmed Newborn-Screened Patients with Pompe Disease across the Disease Spectrum. *Pediatrics* **2017**, *140*, S24–S45. [CrossRef] [PubMed]

14. DeLuca, J.M.; Kearney, M.H.; Norton, S.A.; Arnold, G.L. Parents' experiences of expanded newborn screening evaluations. *Pediatrics* **2011**, *128*, 53–61. [CrossRef] [PubMed]

15. Waisbren, S.E.; Albers, S.; Amato, S.; Ampola, M.; Brewster, T.G.; Demmer, L.; Eaton, R.B.; Greenstein, R.; Korson, M.; Larson, C.; et al. Effect of expanded newborn screening for biochemical genetic disorders on child outcomes and parental stress. *JAMA* **2003**, *290*, 2564–2572. [CrossRef]

16. Herbert, M.; Cope, H.; Li, J.S.; Kishnani, P.S. Severe Cardiac Involvement Is Rare in Patients with Late-Onset Pompe Disease and the Common c.-32-13T>G Variant: Implications for Newborn Screening. *J. Pediatr.* **2018**, *198*, 308–312. [CrossRef]

17. Rairikar, M.V.; Case, L.E.; Bailey, L.A.; Kazi, Z.B.; Desai, A.K.; Berrier, K.L.; Coats, J.; Gandy, R.; Quinones, R.; Kishnani, P.S. Insight into the phenotype of infants with Pompe disease identified by newborn screening with the common c.-32-13T>G "late-onset" GAA variant. *Mol. Genet. Metab.* **2017**, *122*, 99–107. [CrossRef]

18. Tarnopolsky, M.A.; Nilsson, M.I. Nutrition and exercise in Pompe disease. *Ann. Transl. Med.* **2019**, *7*, 282. [CrossRef]

19. Herbert, M.; Case, L.E.; Rairikar, M.; Cope, H.; Bailey, L.; Austin, S.L.; Kishnani, P.S. Early-onset of symptoms and clinical course of Pompe disease associated with the c.-32-13 T > G variant. *Mol. Genet. Metab.* **2019**, *126*, 106–116. [CrossRef]

20. Timmermans, S.; Buchbinder, M. Patients-in-waiting: Living between sickness and health in the genomics era. *J. Health Soc. Behav.* **2010**, *51*, 408–423. [CrossRef]

21. Pruniski, B.; Lisi, E.; Ali, N. Newborn screening for Pompe disease: Impact on families. *J. Inherit. Metab. Dis.* **2018**, *41*, 1189–1203. [CrossRef] [PubMed]

 International Journal of
Neonatal Screening

MDPI

Review

Is Newborn Screening the Ultimate Strategy to Reduce Diagnostic Delays in Pompe Disease? The Parent and Patient Perspective

Raymond Saich *, Renee Brown, Maddy Collicoat, Catherine Jenner, Jenna Primmer, Beverley Clancy, Tarryn Holland and Steven Krinks

Australian Pompe Association Inc., Kellyville, NSW 2155, Australia; Renee@australianpompe.org.au (R.B.); Maddy@australianpompe.org.au (M.C.); Catherine@australianpompe.org.au (C.J.); Jennaprimmer1@bigpond.com (J.P.); Beverley@australianpompe.org.au (B.C.); tarryn_80@hotmail.com (T.H.); stevenkrinks@msn.com (S.K.)
* Correspondence: Raymond@australianpompe.org.au; Tel.: +61-2-9629-2842

Received: 18 December 2019; Accepted: 6 January 2020; Published: 9 January 2020

Abstract: Pompe disease (PD) is a rare, autosomal-recessively inherited deficiency in the enzyme acid α-glucosidase. It is a spectrum disorder; age at symptom onset and rate of deterioration can vary considerably. In affected infants prognosis is poor, such that without treatment most infants die within the first year of life. To lose a baby in their first year of life to a rare disease causes much regret, guilt, and loneliness to parents, family, and friends. To lose a baby needlessly when there is an effective treatment amplifies this sadness. With so little experience of rare disease in the community, once a baby transfers to their home they are subject to a very uncertain and unyielding diagnostic journey while their symptomology progresses and their health deteriorates. With a rare disease like PD, the best opportunity to diagnose a baby is at birth. PD is not yet included in the current newborn screening (NBS) panel in Australia. Should it be? In late 2018 the Australian Pompe Association applied to the Australian Standing committee on Newborn Screening to have PD included. The application was not upheld. Here we provide an overview of the rationale for NBS, drawing on the scientific literature and perspectives from The Australian Pompe Association, its patients and their families. In doing so, we hope to bring a new voice to this very important debate.

Keywords: Pompe disease; newborn screening; diagnosis; infantile onset Pompe disease; late onset Pompe disease; patient perspective

---

1. Introduction

Pompe disease (PD), also known as glycogen storage disease II or acid maltase deficiency, is a rare, progressively debilitating lysosomal storage disorder. It is named after Joannes Cassianus Pompe, who first described a case of idiopathic hypertrophy of the heart in a 7-month old infant in the Netherlands in 1932, noting massive vacuolar glycogen accumulation not only in the heart but in all tissues examined [1].

Affected patients have an autosomal-recessively inherited deficiency in the enzyme acid α-glucosidase (GAA, also called acid maltase). This deficiency leads to accumulation of glycogen in multiple tissues, especially in the skeletal muscles, heart, and liver [2]. These glycogen deposits disrupt muscle cell architecture and function, causing progressive motor, respiratory, and cardiac dysfunction. While both genders are equally affected, PD is a spectrum disorder in which the age at symptom onset and rate of deterioration can vary considerably.

The clinical impact of PD is determined primarily by the amount of residual GAA enzyme activity. Enzyme activity is absent or minimal (<3%) in infantile-onset disease (IOPD), but may be reduced to varying degrees (3%–30%) in those with juvenile-onset (JOPD) or late-onset disease (LOPD) [3]. A lower residual enzyme activity level is associated with earlier onset, more severe disease, faster progression, worse prognosis and a shorter survival time [4]. In IOPD clinical symptoms typically become apparent within the first few months of life; prognosis is poor such that without treatment most infants die within the first year of life [5].

The rarity of PD combined with a variety of overlapping clinical signs and symptoms hamper its diagnosis and the initiation of therapy [6–8]. Such delays have a significant negative impact on patients and their families. Recent research has shown that newborn screening (NBS) appears to be better at identifying PD cases than does clinical examination, especially for classical IOPD [9]. Is NBS the ultimate strategy to reduce diagnostic delays in PD? Here we provide an overview of the rationale for NBS, drawing on the scientific literature and including perspectives from patients and their families. In doing so, we hope to bring a new voice to this very important debate.

2. Diagnostic Delay

Diagnostic delay is common in PD, and it exists across the disease spectrum [8]. Data from the Pompe Registry has found the diagnostic gap to be shortest (average 1.4 months; range: 0.0–13.9 months) in patients with classic IOPD and longest in patients with JOPD (average 12.6 years; range: 0.0–60.0 years). Delays are also significant for LOPD (average 6.0 years; range: 0.0–49.8 years) [8]. Within Australia, IOPD diagnosis can occur within a few months of initial symptom onset, but diagnostic delays of up to 7 months have been reported in the literature [10].

The impact of this delay is such that for many patients, health and functional status is often already severely impaired at the time of their diagnosis. Analysis of data from 53 patients (age range 0–64 years) has shown that at the time of diagnosis [11]:

- Classic IOPD patients—cardiac function, hearing, muscle strength and motor development were all impaired, one in three (36%) required supplemental oxygen and two in three (64%) required nasogastric tube feeding;
- LOPD patients—advanced muscle weakness and impaired respiratory function were present, causing varying degrees of handicap, and respiratory support (14% of adults) and use of a wheelchair (7% of adults) were required.

2.1. Barriers to Timely Diagnosis—Australian Perspectives

Although not specific to PD, prompt diagnosis of rare diseases can have many important positive ramifications. Prompt diagnosis facilitates access to appropriate treatment, it can help parents to better understand their child's condition and explain it to others, it may reduce the burden of blame parents feel and it may alleviate some of the stress of the unknown [12].

Australian research (Table 1) has highlighted a lack of screening tests and limited knowledge amongst healthcare professionals as key barriers to diagnostic delays in rare diseases, with the authors calling for more educational support and wider access to a multi-disciplinary team approach to patient care [13]. Australian pediatric research has shown that more than half of the children with a rare disease were not diagnosed until after referral to a clinical specialist in a large metropolitan pediatric hospital [12], confounding the diagnostic delay.

Table 1. Australian survey data: Diagnostic delays are common in rare diseases.

Age Group	Results	Reference
Adults	Time to diagnosis: • 1 year in 51.2% of cases • ≥5 years in 30.0% of cases Number of doctors seen to get confirmed diagnosis: • 1–2 in 33.7% of cases • 3–5 in 37.4% of cases • ≥6 in 28.8% of cases Number with at least one incorrect diagnosis: • 45.9% of cases	Molster, 2016 [13]
Children	Time to diagnosis: • 1 year in 59.8% of cases • ≥3 years in 8.0% of cases Number of doctors seen to get confirmed diagnosis: • 1–2 in 12.5% of cases • 3–5 in 41.8% of cases • ≥6 in 27.7% of cases Number with at least one incorrect diagnosis: • 27.3% of cases	Zurynski, 2017 [12]

Key Considerations:

- Receiving a diagnosis of a rare disease is a life-changing event; delays in receiving a diagnosis are associated with anxiety, stress, symptomatic worsening, inappropriate use of resources and lack of access to appropriate support and care;
- Health professional education is needed to increase awareness of rare diseases and improve the diagnostic process;
- Resources, including access to multi-disciplinary care teams, are needed to support the requirements of people newly diagnosed with rare diseases.

Specialist referral is reported as a pivotal step in obtaining a clinical diagnosis of rare diseases in Australia [12]. In accordance with this, a recent European survey exploring diagnostic odyssey in PD found that circuitous involvement of several healthcare professionals increased the diagnostic delay of IOPD by 200% compared to direct referral to a specialist center [14].

Australians living in rural and remote areas have additional diagnostic barriers, confounded by lack of proximity and timely access to such specialist healthcare providers. In such cases, the typical care pathway involves multiple steps—initial parental recognition that something is not right with their baby, general practitioner (GP) acknowledgement, and referral to a pediatrician—before the child can be seen in a specialist center. This stepwise approach can take a minimum of 3 months and can place a significant financial and psychosocial burden on the family. Some babies survive the time it takes to confirm a diagnosis, but many do not.

2.2. Diagnostic Delays—Australian IOPD Experiences

In the case of PD, the presenting symptoms are diverse and not likely to be suggestive of this diagnosis unless another family member or relative has already been diagnosed [15]. A high index of clinical suspicion is needed to ensure that patients are appropriately examined and tested. A doctor with intimate experience in the diagnosis or management of PD may be more likely to recognize symptoms in an undiagnosed patient; unfortunately very few doctors will have seen a PD patient in Australia. Testing for PD is a relatively simple procedure that can be undertaken in a few days.

Clinician awareness of, and increased vigilance for, the early symptoms of PD are therefore also important contributory factors to timely diagnosis [10].

A particular case in point is that of NP (Figure 1). NP was born 4–6 weeks prematurely; whilst he developed normally at first, early symptomology became apparent within the first few months of his life. Hypotonia, dysphagia, and developmental delays resulted in the family GP diagnosing failure to thrive at age 2 months and NP was placed on a wait-list for pediatrician referral. At age 6 months, not having seen the pediatrician, NP contracted a viral infection. He was hospitalized initially in a regional hospital 58 km from home and transferred 2 days later to the main tertiary hospital, 168 km away. Upon being admitted to the tertiary hospital blood samples were taken and sent interstate (to South Australia) and overseas (to the USA) for analysis. Ten days later the results were returned and NP was formally diagnosed with IOPD. The extent of physical damage sustained to his body during the time to diagnosis was such that NP was unable to survive, passing away aged 32 months.

<table>
<tr>
<td></td>
<td>

"With the logistical issues from living in a rural area, despite a known family history of PD, it still took more than 7 months to get a diagnosis of IOPD. He never regained movement in his legs and did not fully recover from the viral infection. He always had breathing difficulties and respiratory infections. The heartbreak at the loss of such a young life after undergoing months of uncertainty is difficult to comprehend, particularly given that earlier diagnosis would have enabled him to have access to treatment and potentially provided a better long-term outlook."

NP diagnosed at age 7 months.

</td>
</tr>
<tr>
<td>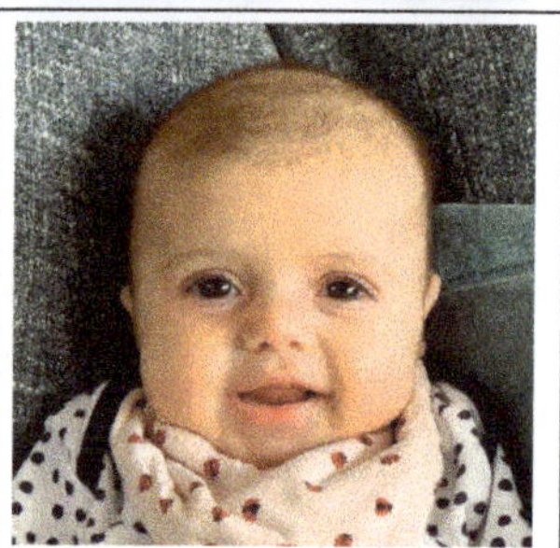</td>
<td>

"She had been a more sleepy baby than our first and we put it down to jaundice but then she was slow to put on weight and fell behind in some key markers at maternal health appointments. The maternal health nurse noticed her shallow breathing at around 9 weeks; I took her hospital they put it down to suspected bronchiolitis and we stayed the night for monitoring. At the next weeks' appointment, the pediatrician picked up that the hospital chest X-ray showed her heart was enlarged. We then saw a pediatric cardiologist who diagnosed her with hypertrophic cardiomyopathy and the concern was to why at a young age she has this and the reasons being life threatening. We then sought further advice from the hospital team and they ran a series of tests after she was admitted at around 12 weeks of age and she never left. We anxiously waited for test results, which validated one of the worst outcomes of Infantile Pompe. It all happened so quickly and she passed away at 14 weeks old, just 99 days. We had never heard of this disease and have no family history on either side of the condition."

LC diagnosed at age 12 weeks.

</td>
</tr>
</table>

Figure 1. Parents' perspectives on Australian infantile-onset disease (IOPD) diagnostic experiences. In lieu of patient informed consent, photographs and comments have been provided by, and reproduced with permission from, the parents of these two children both of whom were deceased at the time the paper was written.

In the last 5 years, the Australian Pompe Association is aware of at least four infants passing away in their first few months of life. These include a baby girl (LC, Figure 1), who was diagnosed at age 12 weeks and three baby boys the youngest of whom died at age 15 weeks. A further two young children, one of whom was diagnosed at age 3 months and died at 19 months and the other who was diagnosed at 9 months and died at 24 months, both had commenced enzyme replacement therapy (ERT) but experienced substantial immune responses [10].

3. What Do We Know about NBS for PD?

The overarching aim of any NBS program is to improve early identification of patients with treatable genetic metabolic disorders diseases in order to confirm diagnosis and initiate management to improve overall health outcomes [16]. Since the initiation of NBS for phenylketonuria, the criteria published by Wilson and Junger in 1968 have provided guiding principles to determine which conditions should be added to the panel [17]. These criteria include availability of a suitable screening test, an effective treatment, an early onset form of the disease that would be debilitating if not treated soon after birth, and consideration of overall cost-effectiveness [18].

Population NBS for PD was first implemented in Taiwan in 2005. A decade later, in 2015, it was added to the USA Department of Health and Human Services-endorsed Recommended Uniform Screening Panel (RUSP) [19]. As of November 2019, implementation has occurred in 23 states and a further 9 states are actively pursuing implementation or conducted pilot studies [20]. Including Taiwan, an estimated 239,333,512 people benefit from the protection of NBS for PD. Pilot NBS studies have also been conducted in Austria, Italy, Hungary, and Japan [21]. Whilst establishment of these programs is challenging and controversial, key driving factors include [19]:

- The development of promising new treatment options;
- Advances in screening technology;
- Advocacy by special interest groups.

3.1. Benefits of NBS for PD

3.1.1. Reduced Diagnostic Odyssey in IOPD

The primary benefit of NBS for PD is the potential to identify patients before symptoms arise, enabling timely initiation of therapy before irreversible damage occurs. Babies with IOPD identified via NBS would be eligible to receive treatment at around 22 days of life, compared to 4 or 5 months of age when relying on symptom-based referral and subsequent diagnosis [8].

Research based on using different decision-analytic models has demonstrated NBS to be superior to clinical examination in identifying IOPD cases, with a significant, positive impact on projected health outcomes [9]. Identifying 40 cases of IOPD via NBS would avert 13 (range 8–19) deaths and 26 (range 20–28) cases of ventilator dependence amongst babies surviving to 36 months of age, assuming all children were treated with ERT. The authors compared their analysis to available real-world data from infants who had undergone NBS for PD in the USA pilot studies at that time, noting that this data supported their models and indicated that the number of cases likely to be detected would be at the upper range of these predictions [9].

3.1.2. Greater Knowledge of Reproductive Risk

At present PD is carried undetected in the community. While patients who have a family history of PD are cognizant of its impact, for many patients there is no such history and a diagnosis of PD can devastate these unexpecting families. NBS provides secondary benefits that positively impact these families, and the wider community, by creating an opportunity to inform about reproductive risks [22]. Early diagnosis of LOPD enables these patients to make better-informed choices regarding future family planning. In addition, a positive NBS screen can be the stepping-stone to identifying reproductive risks in the parents before the birth of a second child. NBS will provide recourse to advice and options, and with knowledge comes informed choices.

Clear frameworks would need to be established to ensure patients and their families are closely followed-up and provided with access to genetic counseling. Equipping key healthcare providers, such as the GP, with information about carrier results and their reproductive implications may help to facilitate parents' understanding of their child's NBS results.

3.1.3. Improved Understanding of the True Prevalence of PD

PD has been thought of as a rare genetic metabolic disease; estimates vary but the literature generally states an incidence of approximately 1 in 40,000 births, of which one-third are infants with IOPD. The Australian Pompe Association membership register does not reflect these IOPD numbers. The youngest treated child currently listed in the Australian Pompe Association membership register is 5 years old; the group is aware of a younger child, aged 4 years, who is diagnosed but not a currently registered member of the Association.

Published Australian data from the mid-1990s estimated an incidence of 1 in 201,000, a prevalence of 1 in 146,000 and a carrier frequency of 1 in 191,000 [23]. No breakdown on the type of PD was provided. As part of reforms into the Government-funded Life Saving Drug Program, a protocol has been established to review the prevalence of PD in Australia based on literature and other published data sources [24]. Data from this review are not yet available, but it is recognized that it will be limited by the availability and incompleteness of identified datasets.

NBS for PD provides a means to help better quantify the true prevalence of this disease. The number of PD cases identified in NBS programs are summarized in Table 2 [21]. These data demonstrate that NBS is finding up to twice the number of cases than were previously thought to exist, underscoring the immense difficulty in diagnosing this condition based on clinical symptomology alone.

Table 2. Results from newborn screening (NBS) programs for Pompe disease (PD). Adapted from Bodamer 2017 [21].

Country and Region	Sample Size	Total Cases of IOPD	Total Cases of LOPD	Prevalence
Taiwan	473,738	9	19	1/16,919
Austria *	34,736	0	4	1/8684
Italy *	3403	0	0	-
Hungary *	40,024	7	2	1/4400
USA (State):				
Illinois *	166,463	2	9	1/15,133
Missouri	269,500	4	20	1/11,229
Washington *	154,544	0	5	1/31,000
New York	390,000	1	30	1/165,000

* Pilot studies.

It is possible that in Australia alone up to two or more babies may die every year without ever having been diagnosed with their underlying IOPD; with these deaths being registered instead as cardiomyopathy or other symptoms of unknown cause. Importantly, the diverse origins of the Australian population may alter the prevalence of PD in the community, but a true understanding of this will not be known without including PD in the national NBS program or, at the very least, investing in a pilot NBS study.

3.2. What Have We Learnt from Current PD NBS Programs?

NBS for PD was first piloted in Taiwan in 2005 and introduced into the Taiwan NBS panel in 2007 [25]. Modification of methodology and systems over time has resulted in a substantial shortening of time to first diagnosis from 19 to 9 days and the time to first treatment initiation from 26 days to 1 day after IOPD confirmation.

Many factors can influence an individual's response to treatment with ERT, including their age and the extent of preexisting pathology [26]. Experience in Taiwan shows that the sooner ERT is commenced, the better the results are [27]. Commencing ERT even a few days earlier can lead to better patient outcomes [28]. For example, 100% of the patients identified through NBS in Taiwan who were initiated on ERT at 6–34 days, and treated for a median of 63 months, remained ventilator-free and have been able to meet age-specific developmental milestones, such as normal independent walking age [29]. By comparison, studies assessing the long-term outcomes of IOPD patients in countries that do not have NBS (including the UK, Germany, the Netherlands, and Italy) report that a substantial number (27%–40%) of children pass away within the first years of life despite ERT [30].

The Taiwanese experience demonstrates that earlier intervention with ERT in IOPD cases can have a positive impact by reducing the future burden of this disease. Moreover, the Erasmus university PD variant database (http://www.pompevariantdatabase.nl/pompe_mutations_list.php?orderby=aMut_ID1) provides data linking the progression and outcomes of over 860 patients with their genetic errors. If genetic sequencing is included as part of NBS confirmatory protocols, this information provides an opportunity to better understand the mutation and to more clearly predict the outcome for the patient.

On this basis, we are of the opinion that the costs of early identification via NBS and early initiation of ERT treatment could be offset by the potential for improved patient quality of life, reduced disease-related disability (such as reduced need for ventilatory support) and reduced associated costs. However, we are unable to support this with hard data at present. Importantly, we remain cognizant that that the current treatment for PD is a first generation product. With considerable research underway for second generation ERT and the potential of gene therapy, the current challenge is to keep Pompe babies as well as possible with the technology available today until a cure is available tomorrow.

3.3. Impact on Immunomodulation Protocols

The development of high and sustained antibody titers (HSAT) is most often associated with IOPD patients who are cross-reactive immunological material negative (CRIM-negative) leading to the recommendation that these patients receive prophylactic immunomodulatory therapy [31]. Whilst ERT has improved clinical outcomes for many patients with PD, there is a risk of developing anti-drug antibodies. HSAT can be associated with worse clinical outcomes. Prophylactic and therapeutic immunomodulation reduce antibody levels, but questions remain as to optimal timing and protocols [32]. HSAT has also been observed in some CRIM-positive patients, raising questions as to how to determine which of these patients should also receive prophylactic immunomodulation [26].

Initiating ERT within the first month of life has not been shown to prevent HSAT [26]. However, experience from Japan, not undertaken in the context of a NBS setting, suggests that early initiation of ERT in the pre-symptomatic period may prevent the progression of IOPD and reduce the likelihood of anti-drug antibody production [33]. These authors suggested that starting ERT before the immune system had matured might have enabled natural immune tolerance. Further research is clearly needed in this area, but this finding opens up possibilities of additional benefits for NBS beyond diagnosis and treatment initiation, because earlier treatment may modify the need for and extent of immunomodulatory approaches.

3.4. Weighing Prognostic Uncertainty against Informed Decision Making

3.4.1. False Positives

Identification of false positive and subsequent prognostic uncertainty is always going to be a core consideration with any NBS program. The extensive Taiwanese experience reports a false positive rate of 0.02% in IOPD targets and 0.01% in IOPD/LOPD targets. Similarly low false positive rates of 0.04% (38.3 per 100,000) and 0.05% (53.2 per 100,000) have been reported from pilot NBS programs in Missouri and Illinois [17] and in a Japanese feasibility study (false positives 0.3%, 2/530) [34].

The literature demonstrates that while prognostic uncertainty does cause heightened anxiety amongst parents in the short-term, there are no documented long-term harms [35]. Options are available to mitigate the issues surrounding identification of false-positives. For example, integrating tandem mass spectrometry with multivariate pattern recognition software to determine which patients warrant second-tier confirmatory testing has been evaluated in the USA with good results, significantly reducing the false-positive rate [36]. The investigators involved in the New York State pilot NBS program suggest the use of second-tier molecular analysis to reduce the burden of referral in screen-positive infants [37]. With their long history of experience in NBS for PD, Taiwanese experts also suggest a second-tier test to reduce the rate of false-positives and facilitate referral of true positives [38].

3.4.2. Early Identification of LOPD

Given that PD is a spectrum disorder, NBS also has the potential for identifying babies with LOPD, noting that their clinical symptomology would not manifest until later in life. Current estimates of the Australian Pompe Association would suggest that if Australia first-tier testing of dry blood spot samples were to take place, for every IOPD case identified there would also be 6 LOPD cases identified. This brings with it several ethical questions surrounding when to start treatment and the burden placed on the patient and their family in terms of waiting for symptoms to appear [39,40].

In the Taiwanese NBS program, 473,738 newborns were screened and 19 LOPD cases had been identified by 2011; 6 (32%) of these patients had commenced ERT between the ages of 1.5–36 months [25]. All 6 patients showed abnormalities with glycogen storage prior to commencing treatment; currently, aged 8–13 years, they have met normal developmental milestones. The remaining 13 patients have not been treated and continue to develop normally, likely representing a milder phenotype [25]. This experience suggests that close monitoring of symptoms and timely ERT initiation, in combination with genetic counseling, education and support should form key aspects in the long-term care of less severe LOPD patients identified through NBS. The ultimate aim being to minimize early medicalization of children while at the same time providing robust protocols to assure the appropriate provision of available treatments.

Aside from treatment and care considerations, there are many positives that must also be considered in the early detection of LOPD cases. These include the ability to learn more about the natural history of the disease and to better As we enter the 2020s equip individuals to make informed choices later in life regarding family planning issues, as discussed above.

The humanistic burden of LOPD is high [41]. Misdiagnosis further impacts this, in terms of costs to the healthcare system incurred as a result of multiple tests and medications being tried without treating the underlying cause and to the patient in terms of ongoing uncertainty, dealing with symptoms and never being quite sure if they would have a better quality life now had they been diagnosed earlier. Early diagnosis of LOPD enables more timely management and may help prevent complications and improve outcomes now that therapy is available [42].

Consider a patient in their 30–40s with mild symptomology. The differential diagnosis for PD is so wide that confirmation of diagnosis may take several years (estimates are in the range of 5–8 years). Importantly anecdotal experience dictates that diagnosis frequently comes after a crisis or other major life event, such as after childbirth, creating additional complexity to what is already a potentially difficult situation.

Without NBS, LOPD patients may go for many years endeavoring to find a diagnosis. The Community loses so much through the cost of unnecessary health care visits while patients struggle for years to find a diagnosis for the want of a simple test at birth that would have alerted the parents or the patient. From a patient's perspective a key advantage to early identification of LOPD via NBS is that it vastly reduces their personal diagnostic journey. In the words of an Australian PD patient, diagnosed in 2010 after a 13-year diagnostic journey:

> *"There is much that needs to be done to help people with rare diseases, particularly around raising awareness to the public and also medical professionals in order for early diagnosis and also correcting misdiagnosis to occur. Had I been diagnosed even in 1997 when I was 17 and received treatment as soon as it became available perhaps my life would be very different today."*

4. Specific Considerations for NBS in the Australian Setting

4.1. Current NBS Policies and Processes

Within Australia, NBS is offered free of charge, and, although not compulsory, participation is high [43]. Australian NBS programs began with screening for phenylketonuria (1967), followed by the addition of congenital hypothyroidism (1977), cystic fibrosis (1981 in New South Wales, 1999 in all other states) and galactosemia (early 1980s). In the late 1990s with advances in technology the list has expanded to include around 25 different disorders [44]. There had been a long silence, with no new conditions added for over 17 years, until the recent initiation of two pilot studies in New South Wales, for primary immune deficiency and spinal muscular atrophy.

All NBS services are coordinated from five centralized screening laboratories (one each in New South Wales, Queensland, South Australia, Victoria, and Western Australia). However, prior to 2018, specific program policies and which conditions to include were individually decided by each state jurisdiction. This has now been replaced with a National Policy Framework, which is accessible via the internet (http://www.cancerscreening.gov.au/internet/screening/publishing.nsf/Content/newborn-bloodspot-screening) [45]. This framework unites these programs for the first time since their inception over 50 years ago. Importantly, it provides a nationally agreed vision and way of working and outlines what will be needed to ensure the ongoing success of the NBS program in Australia; noting that state and territory governments will have the final responsibility for adding the condition in their jurisdictions.

As part of this framework, there is now a national evidence-based process to evaluate proposals to include new conditions in the NBS program [45]. This framework requires the provision of published evidence to support the condition proposed, the test that will be used and the availability and efficacy of treatment such that decisions to include new conditions can be made in line with agreed criteria. These criteria include consideration of (1) whether there is benefit to the baby from early diagnosis of conditions screened, (2) whether the benefit is reasonably balanced against any harms and costs, (3) the availability of a reliable test suitable for newborn bloodspot screening and (4) the availability of a satisfactory system in place to deal with diagnostic testing and follow-up care of babies with abnormal screening results.

The National Policy Framework sets out a step-wise decision-making pathway to carefully evaluate all applications [45]. Following a request to include a new condition in the NBS program and assessment of the available evidence, the possible outcomes include a recommendation to screen, a recommendation to conduct a pilot study, a recommendation to review at a later point in time and a recommendation not to screen.

4.2. Application to Include PD in the NBS

On the basis of the criteria set out in the new NBS National Policy Framework, the Australian Pompe Association took the initiative to submit an application to add IOPD to the NBS program. An initial application was submitted in 2018 but was not reviewed; it was then resubmitted in November 2018 to meet the timelines for a 2019 meeting and decision.

As part of its submission, the Australian Pompe Association calculated that the cost of first-tier dry blood spot testing for PD would be $8.78, amounting to approximately AUD$2.7 million each year, assuming an estimated total birth rate of 310,000 per annum and testing via established tandem mass spectrometry methods. However, recognized additional cost considerations included the need for each State/Territory to purchase a new mass spectrometer because current facilities are believed to be at capacity, funding for additional staff to manage this workload, the establishment and provision of genetic counseling services and costs incurred in second-tier confirmatory testing.

A key consideration of the application was its focus on IOPD, given that this has the most benefit to be gained by minimizing the diagnostic delay. It was hoped that, at the very least, this application would have been seen as a positive step forward and enable a pilot program to be implemented. A pilot program to evaluate first-tier testing has been developed but not implemented due to lack of funding and clinical resources. Such a program would help to define the prevalence of PD in Australia, enable evaluation of the optimal methodology for measuring GAA activity, and better inform costs and resourcing needs for NBS. In addition, it would aid in establishing post-testing diagnostic and confirmatory protocols and procedures for managing JOPD, LOPD, and patients with pseudodeficiency alleles that can lead to false positives.

After consideration by the Standing Committee on Screening, the application has not progressed to a more detailed review. The primary reasons for this being limited longitudinal evidence of survival improvements in treated IOPD patients as a result of identification via NBS and concerns regarding the negative impact of identifying patients at risk of developing LOPD. To a large extent, as has been discussed in this paper, literature providing answers to these concerns is becoming more readily available.

The rapid pace at which new data are emerging, coupled with the increase in uptake of NBS in developed countries like the USA, underscores the drive of the clinical, research and patient communities to provide earlier diagnosis and better outcomes for PD patients and their families. The Australian Pompe Association is encouraged by this and will seek to resubmit its nomination for adding IOPD to the Australian NBS program in the future.

4.3. Access to Current Therapies in Australia

Australia has an established and advanced program for the treatment of rare diseases, the Life Saving Drugs program (LSDP) was established in the mid-1990s and provides people with rare and life-threatening disease with access to medicines that are not listed on the Pharmaceutical Benefits Scheme. Currently 10 conditions are supported by the program, and of the treatments available alglucosidase alfa (Myozyme®) is subsidized for the treatment of IOPD, JOPD, and LOPD.

The LSDP requires that patients meet specific conditions to obtain access to treatment; including initial and ongoing eligibility criteria, and that they undergo annual reviews. Clear protocols are in place for diagnosing the disease and for its ongoing management, including starting and stopping treatment.

The decision by the Australian government to fund treatment through the LSDP is an example to all countries for programs to establish treatment for minority patient groups who face the challenge of living with a rare disease. Treatment is expensive. The only currently approved treatment is ERT and a recent review of the economic costs of PD has established that while available data demonstrate a high cost to patients and healthcare systems, there are substantial gaps in the literature [46]. It is hoped that as new treatment options became available, and competitive interest develops, production methodology will become more cost effective enabling the overall cost of ERT treatment to decline.

4.4. Potential Impact of Future Therapies

Prior to 2006, the only therapy available to patients with PD was palliative. ERT, the only currently available treatment for PD, has been very successful; it can extend the lifespan of babies born with IOPD and stabilize disease progression in patients with LOPD. However, it does not represent a cure. Research continues with many potential avenues including investigations into other therapies such as immune modulation, upregulation of receptor expression, second-generation recombinant ERT, chaperone therapy, substrate reduction therapy, and gene therapy [47]. Recent reviews provide up-to-date information of the available data for these potential new therapies [48,49].

Amongst these therapies, the prospect of gene therapy is of great interest. Currently several biotechnology companies are actively developing gene therapies for PD, while some therapies are still in preclinical development, other have entered early phase clinical trials [48]. Gene therapy has the potential to enable sustained enzyme supply after a single medical intervention; by enabling the patient to produce his or her own enzyme it will vastly change the way in which PD is managed.

As we enter the 2020s, for the first time in 56 years we have an opportunity to significantly reduce the suffering, distress and despair that a diagnosis of PD brings. NBS combined with the potential for gene therapy provides hope that in the not too distant future such patients will be able to say *'Yes I had PD as a baby, but I am fine now thanks to my early detection and treatment'*.

5. Conclusions

NBS has emerged over the past decade as an important contributor to more timely diagnosis and treatment of PD, particularly for babies with IOPD who would otherwise not survive and pass away with their true diagnosis undocumented. Early diagnosis and early access to treatment are pivotal to optimal clinical outcomes.

Australia is not alone in not yet having an NBS program for PD. Here, as in many other countries, the current scenario for patients with PD involves a lengthy diagnostic journey and belated commencement of treatment; in the case of IOPD often after considerable damage has already occurred. NBS facilitates earlier diagnosis and treatment access in a disease in which this timing is absolutely crucial. Identifying and treating IOPD earlier can make a difference between survival and death, between positive outcomes and severe disability. Existing NBS programs have demonstrated the ability to improve patients' lives. Thus, despite its challenges, these positives greatly outweigh the negatives. While we have not had a positive outcome from our application, we hope that by taking the initiative to submit a proposal to include IOPD on the Australian NBS program it will encourage other groups elsewhere to be proactive in investigating and utilizing whatever systems are available in their countries to make similar applications.

The Pompe patient community, both in Australia and around the globe, is highly supportive of NBS [50]. Much headway has been made in ensuring that this patient voice is heard by medical specialists, scientific researchers and industry. Now it is time that this voice is also heard by the regulator to ensure equal and equitable access to NBS and the many benefits it can bring to the lives of PD patients and their families.

Author Contributions: Conceptualization, R.S.; writing—original draft preparation, R.S. with assistance from Hazel Palmer (see below); writing—review and editing, R.S., R.B., M.C., C.J., J.P., B.C., T.H., and S.K.; funding acquisition, R.S. All authors have read and agreed to the published version of the manuscript.

Funding: This research was funded by an unrestricted educational grant from SANOFI-GENZYME AUSTRALIA PTY LTD, grant number SSA-2019-005392 and the APC was funded by SANOFI-GENZYME AUSTRALIA PTY LTD under the same unrestricted educational grant.

Acknowledgments: The authors acknowledge professional writing assistance provided by Hazel Palmer MSc, ISMPP CMPP™ (Scriptix Pty Ltd., Freshwater, NSW, 2096 Australia) in the preparation of this manuscript. The authors thank Bevan Sweerts, Anthony Earp and Jana Chromika, non-author reviewers and employees of SANOFI-GENZYME AUSTRALIA PTY LTD, for their insightful suggestions and comments.

Conflicts of Interest: The authors declare no conflict of interest. The funders had no role in the design of the study; in the collection, analyses, or interpretation of data; in the writing of the manuscript, or in the decision to publish the results.

References

1. Pompe, J.C. Over idiopatische hypertrophie van het hart. *Ned. Tijdschr. Geneeskd.* **1932**, *76*, 304.
2. Angelini, C.; Nascimbeni, A.C.; Semplicini, C. Therapeutic advances in the management of Pompe disease and other metabolic myopathies. *Ther. Adv. Neurol. Disord.* **2013**, *6*, 311–321. [CrossRef] [PubMed]
3. Lim, J.A.; Li, L.; Raben, N. Pompe disease: From pathophysiology to therapy and back again. *Front. Aging Neurosci.* **2014**, *6*, 177. [CrossRef] [PubMed]
4. Herzog, A.; Hartung, R.; Reuser, A.J.; Hermanns, P.; Runz, H.; Karabul, N.; Gökce, S.; Pohlenz, J.; Kampmann, C.; Lampe, C.; et al. A cross-sectional single-centre study on the spectrum of Pompe disease, German patients: Molecular analysis of the GAA gene, manifestation and genotype-phenotype correlations. *Orphanet J. Rare Dis.* **2012**, *7*, 35. [CrossRef] [PubMed]
5. Kemper, A.R.; Comeau, A.M.; Green, N.S.; Goldenberg, A.; Ojodu, J.; Prosser, L.A.; Tanksley, S.; Weinreich, S.; Lam, K.K. *The Condition Review Workgroup. Evidence Report: Newborn Screening for Pompe Disease*; US Department of Health and Human Services: Rockville, MD, USA, 2013. Available online: http://www.hrsa.gov/advisorycommittees/mchbadvisory/heritabledisorders/nominatecondition/reviews/pompereport2013.pdf (accessed on 18 November 2019).

6. Kishnani, P.S.; Steiner, R.D.; Bali, D.; Berger, K.; Byrne, B.J.; Case, L.E. Pompe disease diagnosis and management guideline. *Genet. Med.* **2006**, *8*, 267–288. [CrossRef]

7. Toscano, A.; Montagnese, F.; Musumeci, O. Early is better? A new algorithm for early diagnosis in late onset Pompe disease (LOPD). *Acta Myol.* **2013**, *32*, 78–81.

8. Kishnani, P.S.; Amartino, H.M.; Lindberg, C.; Miller, T.M.; Wilson, A.; Keutzer, J. Timing of diagnosis of patients with Pompe disease: Data from the Pompe registry. *Am. J. Med. Genet. A* **2013**, *161A*, 2431–2443. [CrossRef]

9. Prosser, L.A.; Lam, K.K.; Grosse, S.D.; Casale, M.; Kemper, A.R. Using Decision Analysis to Support Newborn Screening Policy Decisions: A Case Study for Pompe Disease. *MDM Policy Pract.* **2018**, *3*. [CrossRef]

10. Owens, P.; Wong, M.; Bhattacharya, K.; Ellaway, C. Infantile-onset Pompe disease: A case series highlighting early clinical features, spectrum of disease severity and treatment response. *J. Paediatr. Child Health* **2018**, *54*, 1255–1261. [CrossRef]

11. Rigter, T.; Weinreich, S.S.; van El, C.G.; de Vries, J.M.; van Gelder, C.M.; Gungor, D.; Reuser, A.J.J.; Hagemans, M.L.C.; Cornel, M.C.; van der Ploeg, A.T.; et al. Severely impaired health status at diagnosis of Pompe disease: A cross-sectional analysis to explore the potential utility of neonatal screening. *Mol. Genet. Metab.* **2012**, *107*, 448–455. [CrossRef]

12. Zurynski, Y.; Deverell, M.; Dalkeith, T.; Johnson, S.; Christodoulou, J.; Leonard, H.; Elliott, E.J. Australian children living with rare diseases: Experiences of diagnosis and perceived consequences of diagnostic delays. *Orphanet J. Rare Dis.* **2017**, *12*, 68. [CrossRef] [PubMed]

13. Molster, C.; Urwin, D.; Di Pietro, L.; Fookes, M.; Petrie, D.; van der Laan, S.; Dawkins, H. Survey of healthcare experiences of Australian adults living with rare diseases. *Orphanet J. Rare Dis.* **2016**, *11*, 30. [CrossRef] [PubMed]

14. Lagler, F.B.; Moder, A.; Rohrbach, M.; Hennermann, J.; Mengel, E.; Gökce, S.; Hundsberger, T.; Rösler, K.M.; Karabul, N.; Huemer, M.; et al. Extent, impact, and predictors of diagnostic delay in Pompe disease: A combined survey approach to unveil the diagnostic odyssey. *JIMD Rep.* **2019**, *49*, 89–95. [CrossRef] [PubMed]

15. Di Iorio, G.; Cipullo, F.; Stromillo, L.; Sodano, L.; Capone, E.; Farina, O. S1.3 Adult-onset Pompe disease. *Acta Myol.* **2011**, *30*, 200–202.

16. Burton, B.K.; Kronn, D.F.; Hwu, W.-L.; Kishnani, P.S. The Initial Evaluation of Patients After Positive Newborn Screening: Recommended Algorithms Leading to a Confirmed Diagnosis of Pompe Disease. *Pediatrics* **2017**, *140*, S14. [CrossRef] [PubMed]

17. Millington, D.S.; Bali, D. Current state of the art of newborn screening for lysosomal storage disorders. *Int. J. Neonatal Screen.* **2018**, *4*, 24. [CrossRef]

18. Millington, D.S. The Role of Technology in Newborn Screening. *N. C. Med. J.* **2019**, *80*, 49–53. [CrossRef]

19. Baker, M.; Griggs, R.; Byrne, B.; Connolly, A.M.; Finkel, R.; Grajkowska, L.; Haidet-Phillips, A.; Hagerty, L.; Ostrander, R.; Orlando, L.; et al. Maximizing the Benefit of Life-Saving Treatments for Pompe Disease, Spinal Muscular Atrophy, and Duchenne Muscular Dystrophy Through Newborn Screening: Essential Steps. *JAMA Neurol.* **2019**, *76*, 978–983. [CrossRef]

20. Newborn Screening Technical Assistance and Evaluation Program (NewSTEPs). *Disorders Screening Status Map New Steps: US National Newborn Screening Resource Center*; Association of Public Health Laboratories: Silver Spring, MD, USA. Available online: https://www.newsteps.org/resources/newborn-screening-status-all-disorders (accessed on 20 November 2019).

21. Bodamer, O.A.; Scott, C.R.; Giugliani, R. Newborn Screening for Pompe Disease. *Pediatrics* **2017**, *140*, S4. [CrossRef]

22. Bombard, Y.; Miller, F.A.; Hayeems, R.Z.; Avard, D.; Knoppers, B.M. Reconsidering reproductive benefit through newborn screening: A systematic review of guidelines on preconception, prenatal and newborn screening. *Eur. J. Hum. Genet.* **2010**, *18*, 751–760. [CrossRef]

23. Meikle, P.J.; Hopwood, J.J.; Clague, A.E.; Carey, W.F. Prevalence of lysosomal storage disorders. *JAMA* **1999**, *281*, 249–254. [CrossRef] [PubMed]

24. Health Consult. Review of Life Saving Drugs Program Medicines: Pompe Disease. In *Final Review Protocol*; Australian Government Department of Health: Canberra, Australia, 2019. Available online: https://www1.health.gov.au/internet/main/publishing.nsf/Content/E959F2C329B6255ACA258308001F0EE1/$File/Review-Protocol-for-Pompe-disease.pdf (accessed on 20 November 2019).

25. Chien, Y.H.; Hwu, W.L.; Lee, N.C. Newborn screening: Taiwanese experience. *Ann. Transl. Med.* **2019**, *7*, 281. [CrossRef] [PubMed]

26. Desai, A.K.; Kazi, Z.B.; Bali, D.S.; Kishnani, P.S. Characterization of immune response in Cross-Reactive Immunological Material (CRIM)-positive infantile Pompe disease patients treated with enzyme replacement therapy. *Mol. Genet. Metab. Rep.* **2019**, *20*, 100475. [CrossRef] [PubMed]

27. Chien, Y.H.; Hwu, W.L.; Lee, N.C. Pompe disease: Early diagnosis and early treatment make a difference. *Pediatr. Neonatol.* **2013**, *54*, 219–227. [CrossRef]

28. Yang, C.F.; Yang, C.C.; Liao, H.C.; Huang, L.Y.; Chiang, C.C.; Ho, H.C.; Lai, C.J.; Chu, T.H.; Yang, T.F.; Hsu, T.R.; et al. Very Early Treatment for Infantile-Onset Pompe Disease Contributes to Better Outcomes. *J. Pediatr.* **2016**, *169*, 174–180. [CrossRef]

29. Chien, Y.H.; Lee, N.C.; Chen, C.A.; Tsai, F.J.; Tsai, W.H.; Shieh, J.Y.; Huang, H.J.; Hsu, W.C.; Tsai, T.H.; Hwu, W.L.; et al. Long-term prognosis of patients with infantile-onset Pompe disease diagnosed by newborn screening and treated since birth. *J. Pediatr.* **2015**, *166*, 985–991. [CrossRef]

30. Hahn, A.; Schänzer, A. Long-term outcome and unmet needs in infantile-onset Pompe disease. *Ann. Transl. Med.* **2019**, *7*, 283. [CrossRef]

31. Desai, A.K.; Li, C.; Rosenberg, A.S.; Kishnani, P.S. Immunological challenges and approaches to immunomodulation in Pompe disease: A literature review. *Ann. Transl. Med.* **2019**, *7*, 285. [CrossRef]

32. Poelman, E.; Hoogeveen-Westerveld, M.; van den Hout, J.M.P.; Bredius, R.G.M.; Lankester, A.C.; Driessen, G.J.A.; Kamphuis, S.S.M.; Pijnappel, W.W.M.; van der Ploeg, A.T. Effects of immunomodulation in classic infantile Pompe patients with high antibody titers. *Orphanet J. Rare Dis.* **2019**, *14*, 71. [CrossRef]

33. Matsuoka, T.; Miwa, Y.; Tajika, M.; Sawada, M.; Fujimaki, K.; Soga, T.; Tomita, H.; Uemura, S.; Nishino, I.; Fukuda, T.; et al. Divergent clinical outcomes of alpha-glucosidase enzyme replacement therapy in two siblings with infantile-onset Pompe disease treated in the symptomatic or pre-symptomatic state. *Mol. Genet. Metab. Rep.* **2016**, *9*, 98–105. [CrossRef]

34. Oda, E.; Tanaka, T.; Migita, O.; Kosuga, M.; Fukushi, M.; Okumiya, T.; Osawa, M. Newborn screening for Pompe disease in Japan. *Mol. Genet. Metab.* **2011**, *104*, 560–565. [CrossRef] [PubMed]

35. Goldenberg, A.J.; Comeau, A.M.; Grosse, S.D.; Tanksley, S.; Prosser, L.A.; Ojodu, J.; Botkin, J.R.; Kemper, A.R.; Green, N.S. Evaluating Harms in the Assessment of Net Benefit: A Framework for Newborn Screening Condition Review. *Matern. Child Health J.* **2016**, *20*, 693–700. [CrossRef] [PubMed]

36. Minter Baerg, M.M.; Stoway, S.D.; Hart, J.; Mott, L.; Peck, D.S.; Nett, S.L.; Eckerman, J.S.; Lacey, J.M.; Turgeon, C.T.; Gavrilov, D.; et al. Precision newborn screening for lysosomal disorders. *Genet. Med.* **2018**, *20*, 847–854. [CrossRef] [PubMed]

37. Wasserstein, M.P.; Caggana, M.; Bailey, S.M.; Desnick, R.J.; Edelmann, L.; Estrella, L.; Holzman, I.; Kelly, N.R.; Kornreich, R.; Kupchik, S.G.; et al. The New York pilot newborn screening program for lysosomal storage diseases: Report of the First 65,000 Infants. *Genet. Med.* **2019**, *21*, 631–640. [CrossRef] [PubMed]

38. Chiang, S.C.; Chen, P.W.; Hwu, W.L.; Lee, A.J.; Chen, L.C.; Lee, N.C.; Chiou, L.Y.; Chien, Y.H. Performance of the Four-Plex Tandem Mass Spectrometry Lysosomal Storage Disease Newborn Screening Test: The Necessity of Adding a 2nd Tier Test for Pompe Disease. *Int. J. Neonatal Screen.* **2018**, *4*, 41. [CrossRef]

39. Van El, C.G.; Rigter, T.; Reuser, A.J.; van der Ploeg, A.T.; Weinreich, S.S.; Cornel, M.C. Newborn screening for pompe disease? A qualitative study exploring professional views. *BMC Pediatr.* **2014**, *14*, 203. [CrossRef]

40. Wilcken, B. Newborn Screening for Lysosomal Disease: Mission Creep and a Taste of Things to Come? *Int. J. Neonatal Screen.* **2018**, *4*. [CrossRef]

41. Schoser, B.; Bilder, D.A.; Dimmock, D.; Gupta, D.; James, E.S.; Prasad, S. The humanistic burden of Pompe disease: Are there still unmet needs? A systematic review. *BMC Neurol.* **2017**, *17*, 202. [CrossRef]

42. Reardon, K.; McKelvie, P. 090 The expanding clinical phenotype of late onset pompe disease: A multi-system disorder. *JNNP* **2018**, 89. [CrossRef]

43. Royal Australian Collge of General Practitioners. *Genomics in General Practice*; RACGP: East Melbourne, Australia, 2018; Available online: https://www.racgp.org.au/download/Documents/Guidelines/Genomics-in-general-practice.pdf (accessed on 20 November 2019).

44. O'Leary, P.; Maxwell, S. Newborn bloodspot screening policy framework for Australia. *Australas. Med. J.* **2015**, *8*, 292–298. [CrossRef]

45. White, C. *Newborn Bloodspot Screening Working Group: Newborn Bloodspot Screening National Policy Framework*; Australian Government Department of Health: Canberra, Australia, 2018. Available online: http://www.cancerscreening.gov.au/internet/screening/publishing.nsf/Content/newborn-bloodspot-screening (accessed on 5 December 2019).

46. Schoser, B.; Hahn, A.; James, E.; Gupta, D.; Gitlin, M.; Prasad, S. A Systematic Review of the Health Economics of Pompe Disease. *Pharm. Open* **2019**, *3*, 479–493. [CrossRef] [PubMed]

47. Kishnani, P.S.; Beckemeyer, A.A. New therapeutic approaches for Pompe disease: Enzyme replacement therapy and beyond. *Pediatr. Endocrinol. Rev.* **2014**, *12*, 114–124. [PubMed]

48. Ronzitti, G.; Collaud, F.; Laforet, P.; Mingozzi, F. Progress and challenges of gene therapy for Pompe disease. *Ann. Transl. Med.* **2019**, *7*, 287. [CrossRef] [PubMed]

49. Salabarria, S.M.; Nair, J.; Clement, N.; Smith, B.K.; Raben, N.; Fuller, D.D.; Byrne, B.J.; Corti, M. Advancements in AAV-mediated Gene Therapy for Pompe Disease. *J. Neuromuscul. Dis.* **2019**. [CrossRef]

50. House, T.; O'Donnell, K.; Saich, R.; Di Pietro, F.; Broekgaarden, R.; Muir, A.; Schaller, T. The role of patient advocacy organizations in shaping medical research: The Pompe model. *Ann. Transl. Med.* **2019**, *7*, 293. [CrossRef]

MDPI

St. Alban-Anlage 66

4052 Basel

Switzerland

Tel. +41 61 683 77 34

Fax +41 61 302 89 18

www.mdpi.com

International Journal of Neonatal Screening Editorial Office

E-mail: ijns@mdpi.com

www.mdpi.com/journal/ijns

www.ingramcontent.com/pod-product-compliance
Lightning Source LLC
LaVergne TN
LVHW070653170726
843515LV00004B/400